Biomedical Ethics

PREVIOUS ISSUES

Volume 1, 1985	*Ethics, Religion, Education, Welfare, Peace, and the State*
Volume 2, 1986	*The Aesthetics of Humanism*
Volume 3, 1987	*Science and Humanism*
Volume 4, 1988	*Rethinking Humanism: History, Philosophy, Science*
Volume 5, 1989	*The Enlightenment Reconstructed*
Volume 6, 1991	*Meaning in Humanism*
Volume 7, 1992	*Humanism and New Age Thinking*
Volume 8, 1993	*Humanism and Postmodernism*
Volume 9, 1995	*Humanism's Answers: Speaking to the Challenge of Orthodoxy*
Volume 10, 1996	*Living as Humanists*
Volume 11, 1997	*Humanists and Education*
Volume 12, 1998	*Globalization and Humanism*
Volume 13, 1999	*Beyond Reason?*
Volume 14, 2000	*Multiculturalism: Humanist Perspectives* (Prometheus Books)
Volume 15, 2002	*Ecohumanism* (Prometheus Books)
Volume 16, 2006	*The Fate of Democracy* (Prometheus Books)

Subject to availability, back issues of volumes 1-7 can be purchased for $6.00 each, and volumes 8-13 for $8.00 each.

North American Committee for Humanism / Humanist Institute
8014 Olson Memorial Highway
PMB 220
Golden Valley, MN 55427-4712

Fax 763-545-8943
Email manager@humanistinstitute,org

http://www.humanistinstitute.org

Volume 17 of *Humanism Today*

Biomedical Ethics

edited by

Howard B. Radest

*in cooperation with the North American
Committee for Humanism*

 Prometheus Books

59 John Glenn Drive
Amherst, New York 14228-2197

Published 2006 by Prometheus Books

Biomedical Ethics. Copyright © 2006 by the North American Committee for Humanism (NACH). All rights reserved. No part of this publication may be reproduced, stored in a retrieval system, or transmitted in any form or by any means, digital, electronic, mechanical, photocopying, recording, or otherwise, or conveyed via the Internet or a Web site without prior written permission of the publisher, except in the case of brief quotations embodied in critical articles and reviews.

Inquiries should be addressed to
Prometheus Books
59 John Glenn Drive
Amherst, New York 14228–2197
VOICE: 716–691–0133, ext. 207
FAX: 716–564–2711
WWW.PROMETHEUSBOOKS.COM

10 09 08 07 06   5 4 3 2 1

Library of Congress Cataloging-in-Publication Data

ISSN 1058–5966
ISBN-13: 978–1–59102–423–1
ISBN-10: 1–59102–423–4

Printed in the United States of America on acid-free paper

For Information on Leadership Training and Other Programs, please contact:

THE HUMANIST INSTITUTE
2 West 64th Street
New York, NY 10023

Contents

ADDENDUM

Acknowledgments

H oward Radest, founding dean of The Humanist Institute, agreed to edit these proceedings of the annual Colloquium of the Institute's adjunct faculty. His current career has involved him deeply in bioethics. Grateful acknowledgment also goes to the authors for their willingness to permit their essays on a shifted schedule. Responsibility for preparing print-ready version has been mine.

Thanks also to Paul Kurtz for facilitating the publication by Prometheus Books of this series; to Editor-in-Chief Steven L. Mitchell for his wise suggestions; and to my wife, Ana Martinez-Tapp, precision proofreader, ambiguity spotter, and continuing pillar.

Volumes of *Humanism Today* are published by The Humanist Institute, which was founded in 1982 by the North American Committee for Humanism (NACH) as an educational venture to train professional and lay leaders for existing humanist organizations. The guiding principle has been that studying together would enhance all forms of nontheistic humanism, whether they described themselves as religious or secular; Ethical Culturist, Unitarian Universalist, Humanistic Jew; rationalist or freethinker; agnostic or atheist.

More than one hundred students have completed a three-year graduate level curriculum. The Institute's adjunct faculty has gathered annually to consider pressing topics, and this book grows out of the 2003 colloquium. The adjunct faculty met in 2006 to complete a volume exploring viable education undergirdings for endangered democracies and the shaping roles of traditional humanisms.

Robert B. Tapp
Dean Emeritus and
Faculty Chair

Introduction

Humanism and the New Biology

Howard B. Radest

[NOTE: Most of the essays in this book were prepared for the Humanist Institute Faculty Colloquium in the Spring of 2003. Several others have been added since. Berit Brogaard's and Faith Lagay's papers, both of which deal with issues posed by the genome project and the stem cell debate are included with many thanks to them and to *Free Inquiry* where they first appeared [Volume 23, No. 1, Winter 2002-2003]. Two short pieces were written at our request in order to take note of some recent developments: Carmella Epright on the Schiavo case and Kristy Maher on Medicare, Part D. Timeliness is a dilemma in dealing with biomedical ethics. Yet, one hopes to share the ideas and issues that continue to concern us and, by and large, I think we have succeeded. For the rest, we can but try!]

1.

By last count, my bookshelves contained more than 90 volumes on biomedical ethics plus assorted journals. Their content ranges from general texts to personal memoirs. They discuss assisted suicide and in vitro fertilization and everything in between. Some address issues of social policy like access to health care or changes in law. Others are clinical and focus on the ethical dimensions of the cases that are part of the daily work of professional caregivers—physicians, nurses, pharmacists, radiologists, pathologists, social workers, and chaplains. The list of themes and commentaries just keeps on growing! Part of the collection, an expanding part to be sure, deals with the ethics of genetic research and screening, stem cell therapy, and cloning—reproductive and therapeutic. Part reflects debate stirred by White House decisions, or heard in the US Congress, the Council of Europe, and the United Nations. Not least of all, the pulpit weighs in with its voice, often and sadly, too narrow and shrill a voice to be helpful. The collection grows, exhausting shelf space, reading time, and, not least of all my personal budget.

So, I dare a risky question: to what end yet another book on bioethics? After all, the many views that appear in journals and texts, at conferences and consultations already include humanist voices although they are not usually identified with the Humanist movement as such. Humanist values are implicit in caring-for and being cared-by. They are exhibited in the concern for the integrity of patient, family and community that inheres in the work of the health professions. To be sure, these concerns are not always translated into behavior but then, when has any moral ideal been perfectly executed? Echoes of paternalism are still heard despite the ubiquitous reference to "autonomy" and a claimed commitment to what are called "patient rights." Inadequacies of access and treatment are surely present with consequent inequalities. Yet, on all sides, there is real worry about these failures and genuine puzzlement about what to do about them.

It is neither wishful thinking nor sentimentality that leads me to begin by highlighting respect for persons, commitment to human dignity, and attention to distributive justice. These are evident in the discourse and practice of modern health care despite the pressures of scarcity. The past decades have witnessed increasing sensitivity to the personal and the interpersonal in the examining room, the clinic, the hospital and the laboratory, sensitivity to humanistic ethics in short. I say this despite the limits imposed by wrong-headed religious and political beliefs that center on the issue of abortion, that spill over into the debate on assisted reproduction, stem cell research and cloning, and that arise when dealing with assisted suicide and euthanasia. I notice, however, that in just about all of these instances, the operant issue that divides the antagonists is the meaning and interpretation of human worth itself. So I understand that even my opponents' views, wrong as I think they are, are shaped by a concern for humanistic values even if their source and sanction, unlike ours, is otherworldly. So, if these values are already evident in the texts, at work in the field, and in shaping controversy itself, why this book?

The question deserves an answer. To begin, I need to make reference to the institutional setting of these essays. For nearly 25 years, the Humanist Institute has helped to educate leadership, professional and volunteer, for the various Humanist associations in North America. Its faculty—all of whom serve *pro bono* by the way—have met annually to address issues that appear in the public arena, in their work as researchers, teachers, community and congregation leaders, and academics. Suggestions for our agenda come from students in the Institute's classes and from members of Humanist associations. Not least of all in helping us create our agenda is the challenge of a vigorous and often anti-humanist religious fundamentalism and ideological orthodoxy. In response, our goal was and is to create a literature that develops and clarifies Humanist ideas and how they apply to the issues of our age.

In recent years, this goal has led us to respond to the criticisms of post-modernism on left and right and of a re-awakened and often virulent fundamentalism here and abroad. Education, democracy, multiculturalism and environmentalism were self-evidently Humanist concerns. The Colloquium had to re-visit—and will continue to revisit—Humanist philosophy particularly in the light of the contemporary attack on reason, and skepticism if not cynicism about the value of the sciences. We Humanists happily admit that we are "children of the Enlightenment." So it is our privilege and our duty to attend to the Humanist values that guide us—the dignity and worth of the human person, naturalism, rationality, democracy as way of life and not just as political form, the moral and even ontological responsibility of the human being, the sense of appreciation for being, for all being and not just human being. But Humanism cannot afford merely to echo its older truths no matter how precious. The process of Humanism demands, as John Dewey might have put it, continuing reconstruction. And Humanism cannot be content with theory alone. Practice and application inhere in the Humanist notion, as Julian Huxley put it, of evolution become conscious of itself.

We do not pretend that the 16 or so volumes that have emerged so far exhaust the themes we have addressed. Nor do we make such a pretentious claim for the present volume. All are by design works in progress. Similarly, we do not claim that we are the only Humanist voices on the subject. Modern Humanism in the United States boasts a rich literature in journal and text going back nearly 150 years now to rationalism and free thought, Ethical Culture, Emersonian ethical religion, free religion. In the 20th Century, modern Humanism witnessed to the emergence of a Humanist pulpit among Unitarians, the first Humanist Manifesto and later the Second and Third Manifestos, the founding of the American Humanist Association and more recently of Humanistic Judaism and the Council for Secular Humanism. The Humanist Institute plays a special and important role in the project of Humanism. Uniquely in North America, the membership of the Colloquium like the students in the Institute program includes religious and secular Humanists, scientific and Marxist Humanists, pragmatists, rationalists, and idealists, academics, community leaders and activists. The result is not a single authoritative voice but a collection of corroborative voices seeking to orchestrate a Humanist view in its various dimensions. Humanists may issue manifestos, but they do not issue creeds or commandments

2.

Among the endless number of concerns, questions, issues, and problems that both trouble and enrich our experience, why biomedical ethics and why now? My reply is in three parts. The first reason might be called existential. Biomedical ethics is an interesting theme overflowing with questions that are typically humanistic and that are only now in the process of being re-formulated in the light of scientific discoveries and an inclusive democratic polity. Such an environment is particularly apt for Humanists who are by nature a curious and disputatious lot. That curiosity is, after all, central to our commitment to science and free inquiry and to social reform. Given this dynamic tradition, biomedical ethics is a theme that is temperamentally and historically suitable to Humanist style if you will.

A second reason is the need to do our part in shaping the moral agenda of the future. As Michael Werner and David Schafer note in their essays, that future will be shaped in significant ways by the biological sciences. And, as Harvey Sarles reminds us, traditional Western caregiving is properly challenged both by other traditions of health care and novel efforts at cure. Furthermore, contemporary biomedical ethics is an omen of things to come. In its modern incarnation, it is only a few decades old. To be sure, there were earlier soundings and a long rich religious and secular record of care and cure. The appearance of modern biomedical ethics however was stimulated by uniquely modern events, internationally by the horrors of Nazi medical experimentation that led to the adoption of the Nuremberg Code, and nationally, by the immorality of the Tuskegee study, the questionable ethics of human subjects research, and, as Carmela Epright notes, by the cases of Karen Quinlan, Nancy Cruzan and latterly Terri Schiavo.

Of course, medical ethics in the West has a long and interesting history. It goes back to Hippocrates and the Greeks, to Maimonides and St. Thomas in the Middle Ages. Nor is that history irrelevant to the present, a reminder strikingly portrayed by Faith Lagay's resurrection of "humanitas" as a guiding principle in genetic decision-making. But these historic figures and the traditions they represent would hardly recognize the discourse of today's biomedical ethics: the complexities of autonomy, the effectiveness and costliness of treatment, the striking successes of public health and the even more striking challenges which have yet to be seriously addressed, the fascinating outcomes of research which with the advance of the genome project already probe the very being of human being. To these themes, we add the explosive nature of modern biotechnologies, the unknowns of nanotechnology, and the consequent abilities we have both for sustaining and destroying life.

Thus, bioterrorism on the one hand and cure on the other, and in particular cure where even the disease was unknown only a short time ago. Public health measures, medical skill and biotechnology radically reduce infant mortality and extend the human life span. We are enabled to address issues of life's quality, to move from endurance to enjoyment, from survival to expanded powers. Needless to say, all that goes on in clinic and community raises issues of moral complexity and moral puzzlement.

And that brings us to the third part of my response. Humanism has an obligation to address the moral life in all its dimensions. This grows out of its affirmation of human responsibility for the human condition. Whatever over-beliefs individual Humanists may have—and these are many and diverse—Humanists in common accept the fact that no power can save us except ourselves. This is not to say that all problems can be solved or that all threats can be met. The world is friend and enemy all at once and works its will whatever our hopes and desires. Accident, disaster, and tragedy are and will always be part of human experience. But, in so far as situations can be intentionally and intelligently dealt with, it is human beings who must do so.

So, in choosing to address biomedical ethics we are making three statements of Humanist moral concern. First, that the subject is morally exigent and grows more so at an ever more rapid pace. Second, that it is in our power to affect much of what happens in care-giving and public health. We can, in other words, make a difference. And third, that biomedical ethics works in a direct and directive way on the very center of Humanist concern, the human being and his or her quality of living. To realize in fact what human dignity and worth demand in principle requires as its prerequisite a healthy person in a healthy society. This is indeed the note on which Robert Tapp's essay ends the volume. We know how far short we are in realizing that principle. We know too that in any ultimate sense it cannot be realized. But Humanists take evolution seriously and are not perfectionists.

3.

When we met in the Spring of 2003, I did not know quite what to expect of my faculty colleagues as they addressed the deliberately unspecified theme of biomedical ethics. I knew what people in the field—in clinics, government agencies and the like—were talking about. Then too, I teach medical ethics, am deeply involved in clinical consultation, am active in the work of a hospital ethics committee, and in the deliberations of the ethics committee of the South Carolina Medical Association. Finally, I have been at work with public health colleagues on ethical issues raised by preparedness for catastrophic events—natural or man-made. So I came to the Colloquium session expecting further refinement of the issues I was used to dealing with like informed consent, confidentiality, euthanasia, double-effect, living wills, advanced directives, genetic screening, access, and so on. I also came ready to challenge mere familiar rehearsals of familiar Humanist themes. But I was in for a surprise.

At first troubled and gradually appreciative, I heard us address other questions than I'd been prepared for. In a sense, the Colloquium was struggling with more basic questions than the middle range issues of principle, policy, and practice that characterize clinical bioethics. The faculty, in others words, raised questions about what should underlie practice, i.e. they asked about justice, truth, and meaning in a new biological world. In reflecting on my disappointed expectations— and this is in itself an issue of biomedical ethics—it turned out that I was caught in the language of bioethics as an emerging specialty without fully realizing the dangers of its parochialism.

My unwitting absorption in a specialty was in fact challenged by nearly all of the contributions to this volume: by Regal's essay in which he criticizes the adequacy and validity of the bioethical agenda, by Sarles' concern for the body, by Rosenberg's subversion of the notion of bioethics itself, by Schafer call for a crusading science that needed to regain its mission. History as a reminder too in Olds' return to situation ethics and Weldon's reprise of our flirtation with early 20th century eugenics.

So, as I reviewed the texts, I wondered: is it really the case that in a relatively short time—a few decades—biomedical ethics has already narrowed the boundaries of its work, is on the way toward a specialized jargon, etc.? And, more personally, am I so caught up in its work that I am scarcely aware of the assumptions and habits that have become part of my moral reality all unawares? If so, are these good or bad things? In turn, these reflections raise the further question of the mission of biomedical ethics and the yet further question about its domicile in an essentially conservative institution like medicine. Perhaps, caught in the presuppositions, practices, and traditions of medicine, biomedical ethics has undervalued the task of culture critique which is always an ethical obligation. "Know thyself" is, after all, ethics first commandment.

Ironically, these considerations recalled for me the origins of biomedical ethics in religion and philosophy as well as in the major historic moments of moral challenge we have lived through and are living through like World War II, anti-colonialism, internationalism and cultural diversity, human rights and global democratic revolution, an imperative technology, ubiquitous market values, the demography of an aging population. It is certainly true that biomedical ethics—along with other developments in what is called substantive ethics—moved ethics itself from an esoteric study of linguistic usage and meta-ethical formulation to an encounter with the realities of good and evil and the existential dilemmas of birth, dying, and death. In other words, biomedical ethics did not, except derivatively, emerge directly from clinical experience and the puzzles of illness and health, nor did it appear as a species of ethical technicism, i.e. as an "applied" ethics. Its context was the social and cultural problematics of faith, law, justice, and knowledge stirred by the larger historic events of the 20th century.

The essays in this text respond much more to that problematic than to the day-to-day pragmatics of what is now the clinical focus of much of biomedical ethics. Ultimately, of course, a maturing field of inquiry and practice must integrate both culture and clinic and must have its own language and methods. No doubt that process is well on its way. At the same time there are signs of discomfort, as the field must make the effort to respond to cultural diversity, to the impact of genetics, to issues of access.

Summing things up, the essays in the book closest to what may be called the standard issues in the field are my own and Carmela Epright's. Given our respective experiences in clinical settings, this is unsurprising. She focuses on justice, economics and moral conflict in her review of the Lakeberg case involving conjoined twins and the economics of treating them. But, perhaps motivated by the atypical context offered by the Institute faculty, she clearly tries to achieve integration, to connect clinic, case and culture. Yet another type of integration is illustrated in Berit Brogaard's application of philosophic analysis to the "twinning argument" and what it teaches us about the moral status of the embryo. With a different theme and yet with the same integrative intention at work, I address an issue of public policy, namely the democratic conditions for health care rationing, a development that is both inevitable and denied—psychologically, politically, and morally—by policy makers and by the public at large. Kristy Maher's note on "Medicare, Part D" brings that argument up-to-date. Vern Bullough's article on the influence of business values on providing effective and inexpensive contraception instantiates the Humanist tradition of social reform. An interesting and original perspective is offered by Michael Werner's use of the problem of addiction to shed light on the larger theme. So, he sets his analysis of addiction therapies in the context of developing brain studies and against the moralism with which addiction has been and still is treated in public discourse and law.

The essays on the pre-history of bioethics remind us that the field in its original conception was far more imaginative and risky than present practice might suggest. Thus, Mason Olds revisits Joseph Fletcher's "situation ethics" and its relevance to biomedical ethics and Stephen Weldon discusses Herman Muller's work on eugenics. Both celebrate a certain intellectual and moral daring in their subjects whatever one's views of situation ethics and eugenics. Both describe contexts that would threaten the effort of any new field in its attempt to gain both academic and clinical legitimacy.

The scientific and technical concerns of David Schafer and Andreas Rosenberg are skeptical about bioethics itself. Rosenberg, most radically, challenges its very possibility and ends by suggesting a certain pragmatic arbitrariness as the best we can do. Schafer would, as it were, call us back to an Enlightenment sense of the uses of science beyond mere technicism and specialism as a necessary condition for a responsible biomedical ethics. By contrast, Harvey Sarles dares to suggest that scientific medicine may run the risk of deadly ignorance. Knowledge is a plural idea. So-called "non-traditional" medicine, Hindu, Buddhist and Confucian ways of healing for example, and the functions of belief itself ought not to be ruled out of caregiving *a priori*. Indeed, with pressure from patient and public, professional care has been forced to open its boundaries although not always happily. Narrowness, in other words, may arise just because the sciences exist within a particular culture with its particular modes of seeing, knowing, and using. Experience on the other hand always outruns knowledge.

4

As with any book, the question of its uses appears. Clearly, this is not a textbook although the classroom teacher would find it helpful as an auxiliary resource. Read together, the collection opens up questions that ought to be addressed by students as well as by practitioners. Not least of all, both citizens and undoubtedly future patients will benefit by a reading. The former, because policy decisions in health care need a context which does not permit mere expertise to claim authority. The latter, because as we all will inevitably be patients at one time or other we need an ethical literacy that allows us to participate intelligently in cure and care. Beyond functional utility in classroom, clinic, or voting booth, these essays are challenging to the curiosity in all of us.

Ethics is not simply an expert's responsibility. Indeed, as the ethics consultation central to the work of biomedical ethics evolves, it recognizes in practice that all concerned voices and not just the privileged voices of authority have a rightful place at the table. Here, by the way, biomedical ethics is well ahead of ordinary practice and the social habit of delegating away basic responsibilities. At the same time the exercise of responsibility requires intelligence and cannot rely on mere taste or the luck of correct intuitions. In that context, this text can serve the cause of intelligence. Further, as a model of the intersection of personal and professional, and of case and culture, biomedical ethics serves a democratic need. In other words, it has things to teach us as much as it has things to learn.

To these larger tasks, this book is dedicated although in its scope it can only make a small but I trust useful contribution to the larger Humanist project.

Scientific Perspectives
on Bioethics

1.

Twenty-first Century Bioethical Problems
Where Is Bioethics?

Philip Regal

THE BIRTH OF HIGH-TECH GENETICS

C hanges in economic systems have interacted with technological developments throughout history to change people's lives
and values. Knowing this, it was predicted since at least the 1920s that
new developments in biology and medicine could eventually change
human life more profoundly than the industrial revolution.

Aldous Huxley's *Brave New World* (1932) is a familiar literary
landmark that reminds us of the interest in biological futurism early in
the 20th century, and one might also recall the utopian eugenic dreams
that had generated widespread and notorious enthusiasm by the 1930s.
The great biologist Julian Huxley, brother of Aldous, had written a
short story in 1926, "The Tissue Culture King." A scientist uses his
knowledge to serve his own self-interest and secure his position with
a tribe in the jungles of Africa. Huxley prompted the reader to ponder
to what ends the foreseeable power of biology would be used. Would
it be used largely to cater to old tribal dreams and aspirations, or
should we try to figure out better things to do with it?

Meanwhile, the visions of a future dominated by biology involved
especially grand predictions for the anticipated "genetic engineering"

or "biotechnology" at the great private foundations, notably Rocke-feller, and among the scientists they assembled and funded. They set out to do nothing less than create the technical basis for this new chapter in human evolution and economics. The high-tech genetic technologies and the economic/commercial enterprises that were expected to emerge around them would change human society more profoundly than the Industrial Revolution, and would indeed change notions about the definitions of life, nature, and even the human species itself. Their vision, in the spirit of genetic utopia, was com-pletely optimistic, and had no downside.

Then, after decades of intense research, the first genes were spliced together chemically with recombinant DNA techniques in the 1970s. Molecular biologists at the National Academy of Science encouraged government to promote the new technology to private investors and directed agencies to accelerate the development of a publicly funded infrastructure for research and industrialization. Legal, social, and economic incentives were developed to blur the line between basic biology and corporate culture and university scientists were encouraged to become entrepreneurial, and the managers of aca-demic programs in biological and medical research and teaching forced close ties with industry.

Industrial biotechnology was born and its character was shaped as it grew throughout the formative 1980s. By the 1990s the biotech community of entrepreneurial university scientists, life-science corpo-rations, lawyers, and their allies in government agencies had become a major political force in the United States and Europe and indeed globally.

ETHICAL MATTERS

Economic and technological forces historically have tended to change ethical standards of societies. Developments in high tech biology, in basic science, medicine, agriculture, and industry generally have tended to assert pressures on ethical perceptions and standards not simply by working their way into "individual life styles." Nothing like simple "market forces" or consumer demand was asserting the pressures. Rather, pressures from special interests were brought to bear within arenas where, in theory, democratic society as a whole was supposed to have the constitutional power to make choices in policy, funding programs, and law after orderly study and debate. Yet too often the public ethical discussions of these have been intellectually inadequate, detached from the politics of policy formulation, much of which takes place in the shadows and heavily influenced by lobbying, old-boy networks, and expensive public relations campaigns.

"Ethical discussions" have been typically little more than rationalizations about the "greater good." The simplifications, claims and assumptions that the advocates have made and continue to make tend not to be carefully and critically examined. Persuasion and decision are done out of the sight of public scrutiny.

The forced merger between academic biology and medicine and industry in the 1980s, for example, generated many conflicts of interest in government and research and thus raised a number of ethical issues. Could scientists provide the objectivity that society had a right to expect? How would financial involvements impact upon doctor-patient relationships and upon health care systems in general? Could agricultural scientists provide objective technical advice to farmers? When asked how they would deal with such matters, administrators typically tended to give vague answers like, "We will evolve solutions as we go along." "But so much good will come of these changes that whatever price we pay will be worth the costs."

There has been a good deal of frustration with such mergers among involved citizens and scientists who have tried to monitor developments in the biotech/biomedical industries. Personally, I had an opportunity to watch the processes by which policies were developed for nearly 20 years from very close to the centers of power in the United States and Europe. I gained this access during the birth of industrial biotechnology in the critical 1980s. I was invited to dozens of discussions because I was a scientific expert on the technical aspects of risks to the environment and human health from genetic engineering. As a result, my knowledge could help analyze as well the technical feasibility of several types of projects.

I was generally disappointed with any of the ethics discussions that did take place, and even disappointed with the contributions that a developing guild of "professional bioethicists" would make to such discussions, whenever they were invited.

In this essay, I will comment further on the above and on some of the ethical issues that have emerged out of high-tech genetics and related reproductive technologies. It will offer some observations on why such issues have tended and still tend not to get significant attention by mainstream bioethics. And this despite the fact that bioethics grew in the 1990s largely in response to the public's desires to have modern medical and genetic technologies develop only with proper discussion and ethical controls.

It is important to keep in mind that modern biological science is involved not simply in medicine and industry, but in the fate of the biosphere itself. Will we use biological knowledge to contain the human population explosion, embrace ecologically sound agricultural practices, begin to grasp the dangers of the loss of biodiversity, and confront the dangers of unregulated biotechnology and nanotechnology? Or will biological knowledge be generated and used along lines that exacerbate global problems?

THE DEVELOPMENT OF A
PROFESSIONAL GUILD OF BIOETHICISTS

Society, the public, the man in the street, is at least vaguely aware that the stakes with the development and use of biological knowledge are enormously high and wants those who have the power to shape, accelerate or impede to consider the ethical, safety and cost implications of publicly funded or otherwise publicly encouraged programs. As a strategy—partly sincere, often for public relations purposes—those in control of the resources, the broad community of biostakeholders if one likes, thus began to hire "bioethicists" and "biomedical ethicists" so that they could assure society that there were indeed professionals assigned to deal with the public's concerns. The "guild" of bioethicists also could provide the stakeholders with talking points. Thus a professional bioethics subculture became an increasing presence at medical schools, hospitals, and on college campuses.

Of course ethics related to biology had long been discussed by scientists themselves, by essayists and theologians, and in literature. *Arrowsmith* by Sinclair Lewis in 1924 and *The Doctor's Dilemma* by George Bernard Shaw in 1906 come immediately to mind.

But by the late 20th century the informed public was becoming increasingly concerned with the complex ethical challenges emerging from developments in biology and in medicine. New books were continually appearing on the market that explored questions about the social, economic, and ethical implications of the new technologies. This robust market was one sign that the informed public's concerns were likely to continue and even broaden.

Graduates of philosophy departments were often hired to guide the development of the new bioethics programs. Turning the issues over to philosophers might have seemed logical to a public that had little understanding of academic philosophy in the United States, or of the ethical issues that the new scientific developments were likely to raise. In turn, the philosophers naturally set the agendas for institutional and public discourse in terms with which they were familiar. Most had been trained in the post-pragmatic, "analytical" vogues of early and middle 20th century Anglo-American philosophy; increasingly some benefited from development in feminist and narrative philosophy; others from re-discovered Aristotelianism and "virtue" ethics, still others were influenced by post-modernism and neo-pragmatism.

However varied and not surprisingly given its roots in clinical practice, biomedical ethics discourse often came to revolve around discrete cases—or classes of cases—and around the logic involved in reasoning about case studies. Analysis and decisions and the, language, meanings, and categories that grew out of such considerations lent themselves to the use of the traditional tools of philosophers.

But even if some good could come from such deliberations, the question of who is qualified to be an expert on bioethics or ethics generally was hardly raised. The concepts of "the expert" and a specialized profession suggest possession of special authority and competence with regard to the identification of issues, their analysis, communication to and discussion with the public. But is it so that bioethicists are superior in their abilities to identify, analyze, and communicate the challenges of the new biology? Can they really be the watchdogs that society expects?

What Bioethics Programs Might Instead Have Been

Another strategy might have been to hire investigative journalists, sociologists, and/or concerned scientists and physicians to try to understand the career aspirations, economic interests and the qualifications of the important stakeholders, and indeed to identify the key stakeholders. Ethical issues could be based in empirical studies. Philosophers and theologians might then be brought into the inquiry along with others *once the structure of the technological/social institutions, the driving forces for their development, and the human implications became well understood.* and philosophical analysis could realistically be assigned its place among priorities.

Philosophers tend not to be empirical investigators. They tend to break empirical phenomena down into conceptual bits rather than to study them and their functions as elements within larger systems or networks. Their habit is to attend to ethical dilemmas and stylized cases. Their questions, no matter how relevant, have a generality that often does not match experience. They tend to ask questions like: "When does life begin; do stem cells have a soul." They discuss the implications of ethical guidelines such as "do no harm" asking whether or not physicians should condemn a family to anguish by keeping a loved one half alive while they experiment on a terminally ill person. Important as such queries may be to the individuals involved, it takes a different kind of mental effort to try to outline and probe complex political dealings such as the national budget for biomedical research, or patterns of malpractice at various hospitals, or false promises made to Congress about time tables for eliminating cancer or mental illness, or high-tech genetic schemes for eliminating starvation. It takes a different kind of mental effort to go on from there and estimate "downstream" social implications of these, outline the ethical questions that they raise, then attempt to map the landscape of alternative policy responses to the predictable social implications these matters raise, and finally outline the ethical questions they pose. And it takes a different kind of mental effort to map the potential landscape of policy responses to such ethical questions.

These issues need to be approached sociologically, cross-culturally and historically. Unfortunately, too many bioethicists confine themselves to more conventional terms of ethical discourse. This narrow approach fails to clarify the larger cultural context in which events occur and the importance of a larger perspective for democratic self-rule. If cultural context were discussed in comparative terms, bioeothicists might well contribute to a broader discussion of social norms, mores, and taboos and to the aspirations of a democratic, multicultural society.

By focusing on topical issues and clinical cases, and by not expanding on the political and economic forces that all too often shape biological and medical research and practice, bioethicists pose minimal threat to the agendas and careers of the policy makers and administrators of programs in biology and medicine. In that way they may even be said to be complicit in generating the very problems they are intended to solve.

To be sure, there have been bioethical critics from various philosophic and religious perspectives but all too few and all too powerless. Secular language has become the most neutral and practical *lingua franca* for modern pluralistic democracy. Despite the essential role that secular perspectives and language play in pluralistic democracy, large religious factions are legitimate stakeholders in bioethical debates. Ethical matters will necessarily be investigated from religious perspectives. Theological criticisms have brought important basic information together that is relevant even in a secular humanistic context. There have also been critics of bioethicists as a new elite of self-proclaimed experts (for example Smith, Wesley J. 2000. *Culture of Death: The Assault on Medical Ethics in America*. Encounter Books, San Francisco), But however various or valid the approach, dissatisfaction with the state of bioethics seems justified.

Ultimately, however, it would be more useful to explore the issues by examining their network structure and functions. It should be kept in mind that, as a recent phenomenon, bioethicists are not merely self-constructions and they are not merely self-proclaimed experts. They are the human elements of programs that have been constructed by the university, the hospital, the medical school, and the government bureaucratic guilds. In important respects they are *appointed* and *institutionally* proclaimed "experts." Once such a guild has been established, it may well be committed to perpetuating itself . . . or it may not. It is simply to soon to tell. But we do know that individuals and groups, more often than not, can be counted on to try to perpetuate themselves and their values and to protect their interests.

Administrators claim that they have established these programs and appointed the leadership of this guild in response to public concerns, and in a sense this is true. However the public may not always be getting what it thinks it is getting, because the general public's concerns and the special interests that determine the career ladders for administrators do not entirely coincide. Administrators are not for example likely to support teams of investigative reporters who might well begin to question the political structure of the institutions that the administrators administer, or the private sector interests that are affiliated with them. Nor are they likely to appoint Socratic gadflies from among the philosophers.

I have watched these developments from many years in the trenches. But I am not the only one who has been uneasy about the formation of a guild of professional bioethicist stakeholders. Yet the concerns from other scientists, physicians, and sociologists go largely ignored by administrators, the media, and politicians. University of California sociologist John H. Evans, for example, in *Playing God? Human Genetic Engineering and the Rationalization of Public Bioethical Debate* (University of Chicago Press, 2002) has made some of the same observations. The book

> explores the social forces that have led to the thinning out of public debate over human genetic engineering. . . . Disputes over human genetic engineering concern the means for achieving assumed ends, rather than being a healthy debate about the ends themselves. . . . This change in focus occurred as the jurisdiction over the debate shifted from scientists to bioethicists, a change which itself was caused by the rise of the bureaucratic state as the authority in such matters.

Actually I have complained about such things and more. One influential lobbyist responded on several occasions with the same joking rebuttal. "Phil, Bismarck said that there are two things that people should not see being made—sausages and laws." And this line would always bring laughter. On several occasions I complained that particular meetings were trying to make policies without anyone present who really understood the particulars of the issues at hand. The same lobbyist would reply with an ironic quip, "If we aren't the experts, why are we here?" And this line too would always bring laughter. I pressed these matters because I wondered what the response would be if we announced the obvious fact that the purpose of such meetings was often to give the appearance of democratic process. Another unspoken purpose was to test the waters of debate and to work out politically useful language. At some of these meetings it was clear that I was a scientific expert who had been invited largely to lend the appearance of democratic process.

I wonder what those who share this lobbyist's view of government and policy would say about the mainstream bioethics phenomenon. Would they see it in terms of what some management strategists call the promotion of "garbage can issues" to divert attention from the truly important issues that managers have been assigned to push through? (See, for example, Michael D. Cohen and James G. March 1986. *Leadership and Ambiguity: The American College President.* Harvard Business School Press, Boston.)

There is little doubt that bureaucrats indeed are modern courtiers, custodians of and yet users of the system—and not "wise men." Their primary goal is to be successful within the context of their particular career ladders and in the eyes of those above them and below them who have the most influence over their advancement. They are unlikely to be the ones who promote and implement ethical agendas.

A few biologists and physicians have been extremely effective in raising and exploring larger ethical issues related to their sciences. This makes sense because they are closest to the scientific material and the sociological and economic waters in which practitioners must swim. But scientists and physicians are not well poised to see beyond the rhetoric and world-views of their professional groups. And those that do begin to see beyond the self-image of their subculture typically do not have the time to flesh out their insights, and may not be willing to risk their grants, promotions, and standing among their colleagues.

Moreover, the ethics that are often taught to young scientists amount simply to "do your job well, and don't cheat." If a broader vision of ethics is taught it commonly amounts to little more than rehearsed arguments to justify continued funding, a litany about the social good that one's profession supposedly does (and to be sure, often largely really does) in the greater scheme of things. But this is not the same thing as critical self-scrutiny. It is not the same thing as making a thorough inventory of the social effects of the ways in which a research or clinical profession is managed and practiced, and of the rhetorical conventions in and surrounding the profession.

Ethics and the Political Economy: Tricks of the Trade

Political economy is a venerable term that refers to the organization of systems of power at higher social levels, such as the "nation state," and it generally includes monetary/economic institutions and patterns of activity as well as power groups, the system of manpower supply, and even national character as relates to the needs of power.

At high levels of political power, ethical considerations tend to be eclipsed by expediency. What is good and what is right is that which is good and right for *the system*. Thus quite commonly "the end justifies the means." That perspective generates a consequential or teleological ethics. Ethical arguments of those with career investments in maintaining The System tend to be phrased in terms of "the greater good" and/or a utilitarian calculus of the greatest good for the greatest number. Quite often policies are cast to benefit most directly those who are high in *the system,* and are rationalized by a generalized "trickle down" or "supply side" notion, i.e. "a rising tide lifts all boats."

Greater good arguments are easy to construct. Almost anything can and has been justified by using facts selectively. It has even been argued that robbery makes jobs for locksmiths and puts hoarded money back into circulation, and that is true enough. Even before Bernard Mandeville and Adam Smith, however, put greater good arguments into semi-secular views of the world, Augustine (who was not typically a consequential ethicist!) was explaining that God has a way of turning evil into good. And the popes of the 12th century maintained that they could violate the teachings of Christ if their aim was the greater good.

Arguments have been made that all the great advances of civilization can be traced to war, and that wars are akin to exercising the muscles and so necessary for a nation's health. The ends justify the means. And anyone with modest imagination can make up scenarios that will turn any event magically into a means to a good end. This is basically what the teleologists were doing when Voltaire wrote *Candide*.

Greater good arguments must, in my opinion, be used with exquisite care and skepticism, if at all. They have been used to defy world opinion, to sacrifice one's own life, to blow up the Twin Towers, to occupy Iraq. They were used by Saddam Hussein to slaughter Iraqis, by Hitler to round-up and kill Jews, by Jews to subdue and kill Palestinians, by Stalin to slaughter peasants, etc.

A major problem with greater good arguments relates to the fact that words are cheap and so one can get away with superficial insights and false promises. It is easy to make glowing promises about the splendid outcomes of a policy. But history has shown that these promises are seldom kept. Further, they lack accountability. They are commonly to be realized in the future, usually after the persons who make them move on to other jobs or retire or die. People in business and government moreover take it as axiomatic that the public has a short memory, will simply adjust to realities, and that public attention can be shifted to newer issues. In the rare cases where an issue survives, the promise-maker can simply blame the failure of the plan on the other party, in much the same way that each Washington administration blames its economic problems on the policies of the opposing party when last they held office.

The rhetoric of political economy tends often to be cast in superficially scientific or quantitative "cost/benefit" terms. I say "superficially" because history has shown that it can be exceedingly difficult and usually impossible to identify costs and benefits dependably, let alone to quantify them. Often costs and benefits are not of the same material nature and so they cannot be weighed on the same scale. There is the problem of how to compare and evaluate short-term benefits and long-term costs, or vice versa. Often, moreover, the costs can fall upon one party and the benefits upon another. In actual practice there is no careful scientific or quantitative weighing of costs and benefits. Too often it comes down to an arm waving debate over those fears that happen to get to the table, and the dreams of benefits that can be thought up to trump them.

Parenthetically, cost/benefit rhetoric is widely considered to have been a failure as a legitimate intellectual and ethical working concept. Thus there has been a movement, fairly successful in Europe, to abandon it. A "precautionary principle" has been suggested. If valid calculations of costs and benefits cannot be made, then the uncertainty should be resolved in favor of precaution when substantial risks are possible.

Another age-old ploy of both government and industrial bureaucracy is to present a moving target. They will insist that whatever it is that one is concerned about is changing and that the situation is being corrected and is under control. Thus, regulation is always improving since whatever you heard yesterday has changed. "Corruption is now being cleaned up." "We just had a meeting to take care of that." "Yes, that was a problem last year. Our recent reorganization should take care of that. We now have new guidelines in place." "Yes, we are funding research that will address those problems." "We have been looking at that." And so on.

The programs to promote the new biology and medicine are usually argued for in the traditional rhetoric that has long been used to represent the interests of whatever the prevailing political economy happens to be, whether Wall Street capitalist or Stalinist communist or in between.

What ethical values are implied when government decides to spend public funds on programs that have negative effects on health? This happens of course, for example notoriously in the case of government subsidies to tobacco growers and companies. But, as I have said, bioethicists tend to define their field narrowly and do not normally deal with such matters. They might even call it advocacy, i.e. a violation of moral objectivity, rather than ethical analysis if some of their colleagues were to condemn the inhumanity and lethal implications of government subsidies to tobacco growers, or the distribution of cigarettes by the military to troops. It tends to be physicians and public interest groups that raise, explicate, and call for action on such issues.

Yet if the public raises politically serious concerns about an issue, as in the case of stem cell research, then the bioethicists are called upon and respond. One senses almost an attitude of "speak only when spoken to." A cynic might even suspect that administrators would not hire bioethicists who were bringing ethical abuses to light that were not already on the public radar screen. Yet if "ethicists" cannot dig out the "deeper" ethical issues and alert society to these, then wherein is their profession?

Power Philosophies, Court Battles, and Corruption

Bertrand Russell uses the term "power philosophies" in his 1938 book *Power* to refer to arguments that may look superficially like genuine philosophy but that in fact function actually to support power. Analyses of how language, customs, and laws contribute to systems of power have progressed considerably since 1938. We know that the line between ideologies and power philosophies is not clear. Legal arguments and not just moral ones have become as suspect as the legal profession and court system themselves. It is disturbing that the leaders of the biotech/biomedical community predict that the "fairness" of the courts and the "wisdom" of the marketplace will shape the development of the new genetic technologies. Yet, this is how the ethical implications of the new genetic era will be dealt with in the real world.

Considering bioethics from this perspective and on an historical scale, I am very nervous indeed about those who write as though corruption was only a minor problem and that there were more important ethical priorities. At the same time, special interests are playing hardball politics to assure the appointment of sympathetic judges, the creation of convenient laws, and the training of useful lawyers. They have been moving to buy off bureaucrats. They are moving to control global markets.

It must be kept in mind that in the real world the line between ethical violation and political opportunism is often not clear. Is hardball political wheeling and dealing and decision-making in smoke filled rooms ethical in a constitutional democracy? Is it corruption? Is it "business as usual" which is typically conducted behind a façade of democratic processes? Is business as usual ethical or is this aspect of real world politics corrupt?

William "Bill" Kristol, son of Irving Kristol—the "Godfather of neoconservatism"—and Gertrude Himmelfarb, is a central player and major strategist in right wing politics. He argues that politics is a blood sport with no referee. Does this imply a form of philosophical or ideological social Darwinism? Does might make right? From this perspective what is ethically correct and incorrect? What is corruption? The Kristols in turn are among the many influential followers of Leo Strauss. Strauss argued that those with power have a natural right to do whatever is necessary to try to maintain social stability (as they believe it to be). Again, then what is ethical? What is corruption? Is corruption merely the failure of power to preserve itself? This topic is too large to deal with properly here. But I want to point out that to debate ethics as though it were a subject matter that can be isolated experientially, onto-logically or epistemologically from issues of corruption and power politics would help to promote a myth or maintain a façade.

Principles of Biomedical Ethics by Beauchamp and Childress (5th edition), a groundbreaking work in bioethics, was reviewed in the March 27, 2002 issue of *JAMA* [*Journal of the American Medical Association*] by Theodore Fleischer, a South African. He praised the classic text for spelling out the four principles of medical ethics: autonomy, beneficence, nonmaleficence and justice. But he suggests that more should have been written about factors involved in achieving social justice. My thoughts are much in the same spirit as those of the reviewer. The subject matter of biomedical ethics involves forces beyond the door of the laboratory and clinic. I would add that it is critical to keep this in mind at a time when the practice of medicine, the organization of research, the system of the production, and the laws are experiencing enormous and rapid transformations.

21st Century Bioethical Issues that Media Tend to Marginalize

Any list of bioethical issues must necessarily be selective. There are a great many of them and they range from the personal to the institutional and cultural. Hopefully I can simplify the task if I deal next with ethics at the levels of policy and system. This in turn affects all the myriad of ethical issues faced by diverse individuals in idiosyncratic situations with various human interest twists. So the emphasis here will be on issues that are close to the policy level.

RECONFIGURATION INTO
LIFE SCIENCES COMPANIES

The biotech movement has stimulated chemical, seed, food, pharmaceutical companies to reconfigure into "life science companies" and networks of companies. This has raised issues related to the increasing monopoly control of resources and markets. Modern industrial countries have long favored antitrust policies to maintain competition on the theory that this will best serve the public interest. But, most of the large life science conglomerates are multinational and have been playing off national as well as state and local governments against one another to avoid regulation get direct and indirect subsidies, enjoy political influence, etc. This frees the conglomerates from democratic oversight and has made the public's desires for ethical behavior irrelevant. These patterns of reconfiguration and globalization have implications for health care, agriculture, environmental quality, fair wages, and various issues such as property rights.

PHARMACEUTICAL INDUSTRY

The pharmaceutical industry has been growing in political power and has been exerting stronger and stronger pressures on the directions, pace, and independence of medical research and education, clinical practice, and experiments with human subjects. These pressures have appear at both federal and local levels. Our deans at the University of Minnesota, for example, asked the faculty to come up with plans to reorganize the life sciences in order better to serve the pharmaceutical industry. We were told that this is the trend at universities across the country. The administration hired a team of management experts that told the clinical faculties in a large open meeting that they had better get used to the idea that they were now "the workers on the factory floor."

Thus the trend is for health care to be managed at the top not by physicians as such but by graduates of management and accounting schools, and by physician bureaucrats who are quite removed from clinical practice.

One systemic problem is said to be that medical education is suffering at many universities as a result. Another systemic problem with drug company influence is that medical education is biased more and more toward therapy through the use of drugs, and that attention to life-style adjustments and self-help etc. are not getting the attention that they should.

Not enough attention is being given to preventative medicine individually and socially at a time when the need for this is better understood. A related complaint is that public health issues get too little attention. For example, the health implications of poverty, crime, drug addiction, under education etc. are kept off the public radar screen in favor of giving media publicity and institutional attention to the newest medical break-throughs that at best promise to help a relatively few seriously ill people and poster children. Should medical schools and university administrators be active in getting young physicians to think about these issues?

Corporate interests in industrial confidentiality have come into serious conflict with ethical standards such as informed consent and the full disclosure of data to patients who are involved in experiments. The principle of prior informed consent was worked out in international venues at Nuremberg and Helsinki. Now the biotech companies are arguing that they cannot afford to risk having their competitors find out about the progress of their research. So precise and complete information must be withheld from patients and others to prevent leaks of trade secrets. The companies argue that a greater good will result since higher corporate profits mean more investment in research and accelerated research will produce products that the market will then distribute to the needy. Even if one takes the matter in purely pragmatic terms, obviously there are a lot of assumptions here that would need closer study than politicians and bureaucrats tend to have time for.

Tied to globalization these forces have been seeking to impose US models of patent concepts, health care programs, etc. on different cultures and their social, political, and economic systems and ethical norms. These pressures and the ethical conflicts that they produce have been exacerbating international political and economic tensions as well as deteriorating international good will.

The argument is made that huge corporate profits are necessary to fund the research that will eventually help the sick and needy around the world. But drug companies have been keeping prices high even in poor countries. Moreover, drug companies spend more on advertisements, public relations, political lobbying, and on gifts to physicians than on research. Studies have shown that physicians tended to be swayed by such gifts even when they are not aware of it. Too often, they prescribe drugs when not strictly indicated, or prescribe brand-name drugs rather than generics, etc.

The World Bank and the International Monetary Fund and other large banking institutions also indirectly promote the life sciences conglomerates and their agendas (facilities, patents, infrastructure, health care financing) when the bankers demand "structural adjustment" as a condition for loans.

LIFE SCIENCES CONGLOMERATES AND AGRICULTURE

Most of the above political realities pertain to agriculture as well. Farmers in Africa or Latin America may have been cultivating crops and may have been replanting their seeds for thousands of years. But now the banking industry and U.S. Department of State are putting pressure on farmers to plant genetically engineered seeds that they cannot legally replant. Thus subsistence and other farmers are being forced to give up self-sufficiency and to abandon traditional community values with regard to seed ownership and exchange. Pressuring such farmers into the global economy puts them at risk from swings in commodity markets, the vicissitudes of global overproduction of particular commodities, and needing as a result to go into debt or lose their farms, etc.

Studies have predicted that tendencies toward monopoly control of the food industry may well aggravate food supply problems and increase food shortages. Yet the rhetoric of promotion by the life science conglomerates has been that their efforts are necessary to solve hunger and starvation problems in poor countries.

The biotech movement in agriculture includes the promotion of laws to give life science companies maximum control over their seeds and minimum public accountability. For example, Canadian courts have ruled that when Monsanto seeds have appeared on a farmer's property it is his responsibility to detect them and turn them back to Monsanto. How they got there is irrelevant. If corporate lawyers continue to get their way then when pollen from genetically engineered seeds fertilizes a neighbor's crops this will not be legally interpreted as contamination. It will be the affected farmer's responsibility to deal with the problem.

These decisions threaten to put organic farmers out of business . They cannot promise that their crops are not grown from genetically engineered [GE] seed and they cannot sue for liability because their crops are being contaminated. Truly a catch 22.

The question of whether or not GE seeds are dangerous may be irrelevant legally in this kind of case. But there are some cases when GE seeds could be dangerous. For example, with so-called "pharm" crops, which are grown to make drugs or industrial chemicals, the contamination could present very serious health risks. Yet, if the legal agenda progresses as it has thus far, the company that produced the genetic pollution would not be responsible or legally liable. As I write, state legislatures are being lobbied not to pass liability legislation, however "right" this might seem. Life sciences companies threaten to boycott any states where they might be held liable.

I do not imply that there is no health or environmental risk from non-"pharm" genetically engineered crops and animal breeds. But these risks have been discussed elsewhere by others and by me and I could not do justice to that highly technical subject here. Let me only comment that if one accepts the fact that science cannot accurately predict which GE "events" are likely to cause damage to human health, and which not, then when GE foods are sold without labeling, humans are being used as guinea pigs without their permission.

Industry does not like to use the term "guinea pigs" publicly, and currently prefers to say that society is "beta testing" the new technology. But from an ethical point of view the issues are the same whatever we call it. Individuals are being experimented on without their permission. It should be kept in mind that "beta testers" are ordinarily volunteers, and real guinea pigs are used in carefully controlled experiments. Both are very different from throwing an untested product out into society and only reacting if anything bad happens.

PROPERTY AND INTELLECTUAL PROPERTY: BODY TISSUES

A comment should be added about how the legal agenda of the life science companies sees the patenting of human tissues. One example would be a court decision that a man (John Moore) did not have legal rights to a constituent of his own blood. Instead the doctor who discovered its uniqueness and patented it held legal power over its use or disuse. Moore could not, for example, sell his own blood constituent. He could not give it to other researchers to use freely without permission from the researcher who patented it.

There are the obvious ethical issues here. The trust between doctor and patient is seriously threatened when a patient suddenly learns that he has an asset that the state, by issuing a patent, would not allow him to control. This might seem like suddenly learning that a guest had patented your family recipe for pancakes and now you were not allowed even to pass it out freely to your neighbors. But in this case, the state would not merely be invading the bedroom or the kitchen, but the body itself so to speak.

This type of issue has taken on international proportions and has threatened the integrity of science with regard to the human genome project and related drug company activities. Part of the human genome project, for example, has been to collect blood samples from tribal peoples who are known as "Items of Historical Interest" or IHIs. Genomists and drug companies are searching for rare metabolic characteristics that may provide clues to disease and health and to new drugs. The tribes were told that they would be donating blood to advance human knowledge. But they began to realize that "advancing human knowledge" really meant that money was going to be made from their blood in places across the mountains and rivers. The tribes were not going to share in this unless they fought for it. And even if the battle was won, they would still have the problem of monitoring and collecting whatever was due to them. This would be difficult if not impossible without a library of current biomedical journals, a patent attorney, and technical skills.

This behavior has damaged the respect that tribal peoples and those who believe that they are entitled to fair and dignified treatment have had for the scientific establishment. It also raises more general ethical questions like: "Is subterfuge ethical?" "Is withholding information ethical?" "Should scientists be involved in subterfuge and in withholding information from vulnerable people?" The greater good subterfuges that are used to justify unethical scientific behaviors have made people in non-industrialized countries suspicious of science itself and have made it more difficult to get legitimate scientific studies done.

THE INTEGRITY OF SCIENCE

Some people think of science only as a set of procedures for discovering facts, the so-called "scientific method." But science is an historical institution and a moral ideal. Whether we view science as the discovery of some objective reality or as a way of elaborate reality from particular perspectives, it is certainly the case that science has struggled historically and successfully to combat superstition, ignorance, myth, and ideology and pursue reliable knowledge of inorganic and organic nature, including humans.

In our time, the partnership between biology, industry, and politics has become conjugally intimate. The lines between the scientific establishment and the political and industrial establishments are scarcely if at all distinct. Commercial and political rhetoric has increasingly been heard coming out of the mouths of university scientists, most notably since the 1980s. Generations of students have now been trained to think that this is a good thing.

It is difficult these days to find biology majors entering the field because they were inspired by knowledge of Darwin, Curie, and Pasteur who believed that science should be fundamental in to the quest for a world based on truth. More and more students dream of riches, or at least good paying jobs. It is not uncommon to have an entire class of students who have never heard of the Enlightenment, the struggles to establish democratic institutions and to end witch burnings, torture, political domination by Church and State, slavery, the subjugation of women. Our students, most of them, certainly do not understand how scientific ideals and thinking contributed to democratic movements. What then will be the results of this development for society's ethical aspirations?

UNIVERSITY/INDUSTRY/GOVERNMENT RELATIONS

There are yet other ethical issues to be raised. Concerned students from the social sciences and humanities occasionally ask me if it is true that their tuition is being used to subsidize the development of biotech programs that industry wants but for which industry does not want to pay. I don't really know the answer, but I do know that their concerns are real and deserve the most careful investigation. Yet I do not expect bioethicists to investigate such questions. And I ask: should this kind of question be on the bioethical agenda?

This much I also know. The life sciences companies put pressure on university administrators and state legislatures to establish research programs and "incubators" for starting up local biotech companies. Universities and lawmakers have been talked into believing, or at least saying, that this is critical to the economy of the state involved. Consequently, the university pays most of the costs of the construction and the maintenance of physical facilities, faculty hiring, start-up funds, library and computer facilities, etc. Some of the money may be appropriated by the legislature. Sometimes industry can be convinced to donate a token amount. The rest of it must come from one pocket or another in the overall university budget.

Furthermore, these entrepreneurial activities take up much of the time of administrators who could otherwise be working on the curriculum, raising money for teaching, etc. These activities can also affect the kind of faculty that is hired, In turn, this affects the type of instruction and range of courses that are offered and the course of study that students are told will make them into legitimate biologists.

I have encountered instances at my own university where a Dean with connections to the pharmaceutical industry used money from teaching budgets to house and pay for industry-servicing facilities, and other administrators with connections to the pharmaceutical industry covered for him. The funds are difficult to track thoroughly, but probably did not add up to more than a few hundred thousand dollars in this case. Anything so "small" is not likely to impact on the tuition rates for tens of thousands of students. But then, we don't know if this was or was not the tip of an iceberg.

One area that needs close study relates to how state legislatures (at public universities) debate the university budget. To be sure, they use an "educational" rhetoric. But they expect the administration to pay for industry-aimed programs out of this "educational" budget, including biotech programs.

When there are budget cuts, industry-aimed programs are protected because of their political backing and cuts have to be made up in large part by tuition increases. I have suggested that we have two university budgets, one supporting education, and the other clearly supporting industry-aiding programs and services (which are surely legitimate at a land grant university). But I am told that this would not be politically feasible. Subsidies to industry would be transparent and open to public debate.

My point in raising matters of administration, tuition, etc. is that they have considerable ethical implications for bioethics as for education and science generally. Once again, I do not expect professional ethicists to add these items to their agenda. Once again, a certain narrowness of focus may be involved..

Ethics and the coming of the post-human

Many of us have taught, written, and spoken about the equity and ethical problems that will be raised by modern reproductive technologies. It is obvious that the challenges are enormous. Indeed many in the biotech community fear opening these issues to full public discussion. They even argue, I think unrealistically, that this could kill the industry and rob society of its enormous potential benefits. In effect, they argue that full ethical discussion could be unethical. Be that as it may, I bring this up as background for mentioning the advocates of the "post-human" who urge that we ignore ethics completely.

There are molecular biologists and others who argue aggressively that GE should be applied to humans. Individual superior genes will be put up for sale and the invisible hand of the market place in its wisdom will eventually create a superior class of beings. [For example, type Extropy Institute into a search engine such as Google or go to http://www.extropy.org/ideas/journal/previous/2002/08-02.html which is *Extropy: The Journal of Transhuman Solutions*. There are other groups as well and they are not all isomorphic. Enjoy!]

These genetically engineered "post-humans" will be so much more intelligent than mere humans that they will do a much better job of developing ethical systems than we can hope to do. The mere humans of today should not stand in the way of ethical progress by raising their petty ethical issues that arise from human primitive biology and inferior intellect. The most ethical thing that we can be doing today is to suspend making ethical judgments altogether and promote the most rapid development of biotech as possible. A truly intelligent ethics will only become possible in the future. One recalls where Nietzsche in *The Genealogy of Morals* argued that morality must not be allowed to stand in the way of the creation of the superman.

> What if a symptom of regression lurked in the "good," likewise a danger, a seduction, a poison, a narcotic, through which the present lived at the expense of the future? Perhaps more comfortably, less dangerously, but at the same time in a meaner style, more basely?— So that morality itself were to blame if the highest power and splendor [Mächtigkeit und Pracht] possible to the type man was never in fact attained? So that morality itself was the danger of dangers? (GM Pref:6; cf. BT Attempt:5)

I would argue—but not in this essay—that these expectations are utterly unrealistic technically even if we were to believe the unrealistic hype that we have become accustomed to hearing from the biotech industry. These expectations are metaphysically naive and biologically unrealistic. They exhibit an astonishing faith in inevitable evolutionary progress driven by "natural selection" and the invisible hand. Inevitable evolutionary progress may be a popular belief. The fact is, however, that most species do not adapt, but go extinct. Recall too that Darwinian evolution has evolved Irish elk, blind cave fishes, and tapeworms—evolutionary dead ends and simplifications.

But technical and biological unrealism are not the point. Our concern here is a widely held general attitude that the advocates for the post-human merely illustrate, albeit in their own strange way. It is the attitude that the benefits of the new genetic technologies will necessarily be so enormous that any ethical judgments that could slow the biotech industry down should be kept off the table.

MENSCH UND ÜBERMENSCH

People tend to forget that an ideology can have enormous social reverberations and ethical implications. Hitler never managed to produce a single Aryan superman with his seductive eugenics programs. Nevertheless do not ignore the profound consequences of such visions. It may be tempting to assume that the creation of the post-human is merely a pipe dream and to ignore the ideologues of this vision of the future. Or it might be tempting simply to debate the details of their vision as a kind of moral game. For example, would it be ethically right to have humanity ruled by a race of superior post-humans? Or should we experiment on the gene pool for a greater good?

But while approaching the issue in these ways might be interesting and even important, it will not necessarily lead to understanding of the potential socioeconomic implications of such an ideology. It has effects whether simply as a way of thinking or as a worldview, a pivot point for *spin*, a banner for mobilizing political action, economic resources, and public support, a hook for directing public discourse, etc.

So we need to find out how these ideas are circulating through the scientific, economic, and political corridors of power, and what effects they are having on the opinion makers and trend setters. And further, what "trickle-down" characteristics are they nurturing?

This again is a case where bioethics is very much at a disadvantage because of its tendency to focus narrowly on case studies and to boil down complex issues into matters tractable in terms of logic, language, and law. What ethical watchdogs should do in the case of post-human agendas is explore the ideological side of issues and decisively expose it to sociological, political, and historical critique.

CODA

It is not my place to offer the growing community of bioethics programs a recipe for reform that they could actually work with, given their own sociology, disciplinary training, and sources of support. Perhaps they could try to find critical and thoughtful scientists and sociologists who would make them uncomfortable while collaborating with them. But of course this is much easier said than done, to say the least.

It seems more likely that the most effective watchdogs will come from outside of the bioethics programs. Scientists and clinicians who can provide critical insights into the socioeconomic and ideological systems in which they have functioned can do great public service here. Sociological studies of the new life sciences should also be encouraged, so long as the practitioners can resist pressures to conform from government, foundations, and university administrators. It should be remembered that all of these—critical scientists, clinicians, sociologists, as well as public intellectuals—are also bioethicists whether or not they have institutional sanction or title.

2.

Bioscience and the Recovery of Nerve

David Schafer

I am under the impression that it is sometimes not useless to place a certain noble trust in one's own powers. A confidence of this sort gives life to all our efforts and lends them a certain verve, highly conducive to the investigation of the truth.
—Immanuel Kant, *Gedanken von der wahren Schätzung der lebendigen Kräfte*, quoted by Ernst Cassirer

SCIENCE AND ETHICS

Most ethical philosophers, at least since the writings of David Hume, have maintained that ethical statements cannot be reduced to statements of fact, and vice versa. But Hume was very much aware that the "is" and the "ought," facts and values, impinge upon each other. An especially noteworthy example is the dependence of ethical judgments on judgments of what is actual and what is possible in the natural world—in other words, on facts revealed by scientific investigation.

In this essay I will argue that ethical arguments in general, and bioethical arguments in particular, necessarily evolve along with our evolving biological sciences, and therefore that they require constant reexamination from this perspective. Moreover, I will argue that the very immaturity of the biological sciences, especially the behavioral sciences, makes it likely that many classical ethical judgments have been, at the least, premature or inexact, or finally downright wrong.

Ethical judgments in general depend on biological sciences at two essential points, and bioethical judgments at three. To begin with, all ethics rests on a model of the human agent responsible for moral judgment and action, and the predominant model has only recently become a scientific one. In order to decide what actions are good in a given situation, one must ask first what actions are genuinely possible, i.e. actually available, to the agent, both "physically" and behaviorally. As is said, "ought implies can."

In the second place, consequentialist ethics requires a knowledge of the probable direct and indirect effects of an action on its human or other living recipients. To be ethically useful, such knowledge must also enable us to assess whether the effects were intended or unintended, beneficial or harmful, or neutral. Even deontological ethics often assesses effects too although it does so implicitly in its choice of the actions it rules out (e.g., murder, robbery, lying, withdrawal of life support). Such knowledge, again, implies a valid model of the living recipient in a variety of actual or possible circumstances.

Finally, bioethics introduces a third point of contact with biological science: the living or biologically active object or process employed as a means of producing an effect, either beneficial or harmful. For instance, it is a common observation that a means found to be beneficial to a recipient in one situation may be harmful to the same recipient in another situation, or to a second individual as a by-product of its beneficial effect on the first. So the science of biology can influence bioethical decisions by altering our understanding of (1) the human agent, (2) the human or other living recipient of a biological action; or (3) the biological means of that action,—or some combination of these.

BIORESEARCH AND BIOTECHNOLOGY

Today the scope of bioethical discussions includes a vast range of ethical issues associated with biological experimentation and technological innovation. Efforts to introduce new commercial biotechnologies entail large, even vast amounts of investment. In turn, this may bias interested parties for or against them. The annual "BIO" conferences that draw representatives from many biotechnological industries illustrate this point. In June, for example, the "BIO2003" conference in Washington, D.C., identified four main themes: (1) national security, (2) health, (3) jobs, and (4) feeding a hungry world. The first of these was emphasized in a featured speech by President George W. Bush, promoting his pending "Project Shield" legislation. It was not lost on observers that the need to use biotechnology for national security implied its prior use to threaten that security.[1]

The potential ethical dimensions of bioresearch and biotechnology often include public and individual safety in both the short- and long-term, cost, morbidity, pain, death, and on occasion catastrophically widespread ecological changes as well. Estimates of such ethically weighted consequences of actions in biological systems inevitably rest on calculations guided by previous research, which may be limited in scope.

Bioethical issues arise in such radically diverse areas as plant and animal agriculture and ecology; human reproduction and the manipulation of human embryos; the use of humans and other animals as research objects; hybridization of dissimilar living strains and chimerization of distinct species are genetically mixed; development of complex prostheses and cross-species transplants; indefinite prolongation of human life; and so on. This diverse and complex agenda places great burdens on the profession of bioethics, since it is not reasonable to expect that any one bioethicist will be fully competent in the science underlying all, or even many, of these areas. A consequence is that there is frequently a danger of oversimplification.

Central to many of the more novel bioethical controversies are the emerging sciences of genomics and proteomics (see below) with their associated biotechnologies and their implied potential to modify the very characteristics of human nature, the natures of other species, and the relationships among them. Anyone with a little imagination can readily envision "doomsday scenarios," sometimes involving unintended consequences, which might enormously complicate the task of rational decision-making.

Just about all issues facing bioethics are of interest to Humanists, but some are particularly so. The latter raise profound questions about *who or what we and other living things are or might become*. Traditional concepts of nature and human nature are pervasive in all societies. However, not only have these concepts not been based on scientific knowledge but frequently they are incompatible with it. In fact, modern biology is such a young science, that it is not surprising that many Humanists let alone others have not yet fully assimilated the implications of its concepts into their own ethical thought, and this despite the fact that some modern biological ideas have been developing for more than a century and a half, i.e. since Darwin. In order to grasp the reasons for this situation, the following brief chronology is intended to suggest the historical magnitude and speed of changes in this field.

BIOLOGY, A YOUNG SCIENCE

David Hume died in 1776, Jean-Jacques Rousseau and Voltaire two years later. The death of Immanuel Kant in 1804 is sometimes taken as the end of the Enlightenment, with its great emphasis on human freedom and scientific inquiry. Enlightenment values arising from the work of these and other modernists are considered by many Humanists to be the essential values of Humanism as well. Yet if a reliable ethics depends on a reliable model of the real world, populated by real human beings, a sound scientific basis for sound ethics had not yet been established in the eighteenth century. Science, particularly the science of life, was still in its infancy, by any standard. There were, of course, scattered bursts of illumination like William Harvey's theory of the circulation of the blood (1628), Niels Stensen's discovery of the parotid duct (1660), Rev. Stephen Hales' measurement in 1733 of the arterial blood pressure of a mare (dramatically, the blood rose in a cannula to a height of 8 feet, 3 inches), and Alessandro Volta's demonstration that frog muscle contracts in response to electrical stimulation

(1792). But, what we now know as human physiology had scarcely even been imagined.

Before the advent of experimental methods adequate to explain life, philosophers had recourse to many speculative theories. One of the most widespread of these was "vitalism," the concept that life requires an immaterial principle or "vital force." This vital force, it was believed, was necessary for the synthesis of all the carbon-based "organic" materials so distinctive of the world of the living. However, the successful laboratory synthesis of a single one of these compounds, urea, proved to be the beginning of the end of vitalism. An adult human produces an average of around 30 gm of urea every day from nitrogen in the tissues and the diet. Urea is excreted into the urine, where its breakdown by bacteria produces the recognizable odor of ammonia. Even this small molecule (molecular weight 60) was imagined on vitalistic grounds to be intrinsically impossible to reproduce outside a living organism. It was not until 24 years after Kant's death, that Friedrich Wöhler synthesized urea from ammonium isocyanate. In the vernacular of today Wöhler might be said to have been "playing God." To scientists and Humanists this was an important early demonstration that we ourselves are part of the natural universe, and perhaps subject to physical laws.

About ten years later, the botanist Matthias Schleiden and the zoologist Theodor Schwann published theories suggesting that all macroscopic plants and animals are composed of microscopic living cells not visible to the naked eye. These cells, it was thought, were held together in higher living organisms by nonliving i.e. "dead," structures. Visionaries of the East and the West had speculated for centuries about the possible existence of what we would today call "superorganisms," made up of lesser organisms. For instance, the Italian philosopher Giordano Bruno had entertained radical ideas of this sort. Born in 1548, Bruno was burned at the stake by the Italian Inquisition in 1600 (the reasons are still controversial). The cell theory of Schleiden and Schwann was, however, the first scientific suggestion that a human being might be usefully regarded as a superorganism, a highly complex "colony" of organisms not individually visible to the unaided eye!

The structural and behavioral similarities, sometimes extremely close, among different kinds of plants and among different kinds of animals had presumably been obvious even to the earliest observers. Often these were the subjects of countless cautionary stories in which animal or plant characters served as surrogates for human beings. St. Francis' reverence for his animate and inanimate! "brothers" and "sisters" typified this awareness of intimate human relationships to the rest of nature. Indeed, this awareness had actually been incorporated by poets and theologians over the centuries into a hierarchical "Great Chain of Being," in which humans had their place as "a little lower than the angels." Yet the refusal to accept this metaphorical notion of universal kinship as literally true was widespread and profound. It may have been one major reason why Charles Darwin himself hesitated for more than a decade before publishing his *Origin of Species* until 1859, and this only after Alfred Russel Wallace had inadvertently nudged him into doing by publishing his own account of evolution (the two accounts were first published together in 1958).

In the late nineteenth and early twentieth century, as the infant sciences of biochemistry and biophysics began to develop, significant research started to generate knowledge of the structures and functions of individual cell types within larger organisms, as well as the nature of the mechanisms by which these functions might be coordinated. In the 1890s Jacques Loeb, one of the founders of the new science of "general physiology," was starting to look closely at common features of cell and organ physiology through the plant and animal kingdoms. The vague and mysterious term "protoplasm" began to acquire specific chemical and physical characteristics. Loeb's book *The Mechanistic Conception of Life* (1912) became an early classic of the detailed mechanistic view of biology. During the same period a surgeon, Alexis Carrel, demonstrated how individual organ tissues could be maintained alive indefinitely out of the body.

At the turn of the century Sir Charles Sherrington in England and Ivan Pavlov in Russia achieved eminence in the study of unconditioned and conditioned reflexes, respectively. Their research advanced our understanding of how far unconscious processes controlled the workings of the nervous system in integrating the functions of cells and in the way we learn and the way we remember. Around 1921 Otto Loewi demonstrated conclusively that the vagus nerve slowed the heart beat by releasing a chemical, acetylcholine. Pavlov was awarded the Nobel Prize in 1904, Carrel in 1912, Sherrington in 1932, and Loewi in 1936.

A REVOLUTION IN PHYSICS

Meanwhile, in another part of the forest, physics was undergoing its own deep upheavals, with Albert Einstein's development of special (1905) and general (1916) relativity and, in the mid-1920s, equivalent formulations of quantum mechanics by Werner Heisenberg, Erwin Schrödinger, and Paul Dirac. Einstein received the Nobel prize for physics in 1921, and the other three in 1931 and 1932. Their major refinements of the dynamics of systems at very high energies or very small distances, respectively, seemed to some, especially among those least hospitable to science, to bring the entire scientific enterprise into question. Einstein, for example, was never comfortable with the probabilistic interpretation of quantum mechanics, commenting that "God does not play dice with the universe." Even Schrödinger, lecturing in the United States in 1927 on his wave model of quantum mechanics, was reluctant to accept the practical implications of his own model when Clinton Davisson and Lester Germer of Bell Labs reported to him excitedly that they had actually *observed* wave-like diffraction of the electron!

Particular attention was given by some philosophic interpreters to a quantum mechanical discovery first enunciated by Werner Heisenberg, now known as "Heisenberg's Uncertainty Principle." This principle states that any two quantities related to each other in a certain way—technically, "canonically conjugate dynamical variables"—cannot both be known with complete accuracy. The minimum errors in these quantities *must* vary reciprocally. Their product, which must have the same dimensions as energy times time, is so small it is not ordinarily detectable on a macroscopic scale. As an example, if a particle having a mass of one-billionth of a gram (1 g = 1/453 pounds) is traveling in a straight line with a speed that is known to within one-billionth of a centimeter (2.54 cm = 1 inch) per second, the location of the particle on that line cannot be determined more precisely than to about one-billionth of a centimeter. On an atomic or especially a sub-atomic scale, an effect of this magnitude certainly cannot be ignored, but its effect on entities the size of cells or even biomolecules is ordinarily negligible. I apologize for the technical detail I've used in describing this effect. Unfortunately, some commentators, especially among those who are not mathematically minded, have tended to give the "uncertainty principle" a crude metaphysical meaning it does not have, e.g. as demonstrating "free will" or else along the lines of "If you know one thing, you can't know something else." Physicists, on the other hand, were aided rather than daunted by Heisenberg's Principle. With it, they were enabled to make new discoveries of sub-atomic cause-effect relationships at ever-increasing rates.

LIVING ORGANISMS AS MACHINES

In this sketchiest of chronologies, I have mentioned only a very few events useful for my present purposes. My next point is that by 1930 tales of cumulative advances in scientific knowledge of the real world, ranging from particle physics to molecular physiology to neuro-sciences to evolutionary biology to geology to astronomy, had reached the ears of those about to draft, sign, and publish a Humanist Manifesto (1933). The Manifesto was, in fact, predicated in large part on *the widely acknowledged success of the scientific method in answering questions about nature formerly left to traditional religion.* Its 34 signers included some 17 Unitarian ministers and one physiologist, Prof. Anton J. ("Ajax") Carlson, of the University of Chicago. The latter was, at the same time, working on a new textbook of physiology that would become one of the most widely acclaimed in the field, *The Machinery of the Body* (1937).

The significance of Carlson's title is profound. It does not just include the obvious mechanical systems in the body, like the musculoskeletal system and the propulsive machinery of the circulatory and gastrointestinal systems. It is inclusive and refers to the various ways all the systems of the body work, including, of course, the brain. To be sure, the fact that the machinery of the body is constructed of meat, bone, and quite a bit of water makes it somewhat different from our usual understanding of "machinery." A further distinctive feature of this kind of machine is its dynamic life cycle, starting from a single multiplying cell, developing to maturity, and ultimately dying. Within limits it is self-developing, self-organizing, and self-maintaining. Such machines are *organisms*, but they are machines nonetheless. This sort of machinery runs on chemical reactions powered by energy retrieved from foods and used by the body's cells to perform both mechanical work and the work of chemical transformations. To learn how this is done requires a detailed understanding of biochemistry, especially the chemistry of the proteins and nucleic acids, like DNA. *All living things are biochemical machines, and we human beings are no exception. In this context, and in the light of my opening paragraphs, we may now accurately use the term* bioethics *to refer to the ethics of, by, and for such machines.*

By contrast, when in October 1996, the Roman Catholic church finally accepted the notion that evolution is a "serious hypothesis" of human development, it did so with a major proviso: "If the human body has its origin in living material which preexists it, the spiritual soul is immediately created by God." In this view evolution alone "is incapable of establishing the dignity of man." In the same way, the Church has continued to insist on an instant of "ensoulment" of the embryo. As a practical ethical consequence, the intransigence of the Church on many reproductive issues, and especially on those related to abortion and embryonic research, might lessen or even disappear if it were not for this absolutist position on the immaterial soul.[2]

REDUCTIONISM

But, we usually ask, is not the whole in this case somehow "greater than" the sum of its parts? Here the word "sum" is not appropriate; "assembly" would be better, and "organized assembly" better still. "Addition" in other words is not the only way of thinking about how parts generate a whole. Biologists often use the sometimes misunderstood reductionist method to determine how complex organisms work. The analogy of the parts of a car or airplane is a good one. Imagine all these parts spread out on the floor of a garage or hangar. When enough is known about how each part works by itself, it is almost always easier to figure out how it will work later when properly installed in the organism. So much for *methodological reductionism*.[3] What is missing when the parts are separated is the proper *organization* of the whole. A holistic representation of the organism's functioning can be obtained, at least approximately, by mathematical synthesis of the integrated functions of the parts.

Daniel Dennett, a reductionist himself, speaks critically of "greedy" reductionism. We know that more complex systems can be explained in terms of less complex ones, e.g., atoms are composed of elementary particles and their properties can be explained in terms of their component particles; molecules are made of atoms, and their properties can be explained in terms of their constituent atoms; living cells are built of many molecules; superorganisms (like humans) are composed of many living cells, and so on. Dennett uses the term "greedy" reductionism to mean the misguided attempt to skip intermediate stages of complexity in an explanation—for an extreme example, to try to explain the song of a bird in terms of elementary particle physics. Even though we may know they are ultimately connected, the effort to do this is a complete waste of time.

There is also ontological reductionism. The philosopher of science Herbert Feigl, originally a member of the Vienna Circle who spent the last decades of his career at the University of Minnesota, and who was a Humanist signer of Humanist Manifesto II (1973), put all philosophies into three types: the philosophy of "nothing but," by which he meant "the reductive fallacies of narrow minded positivism"; the philosophy of "something more," his term for "the seductive fallacies of metaphysics"; and an intermediate philosophy of "what's what," which he explained as "an attitude of reconstruction." To say that an animal is "something more" than a machine is to imply, as vitalism did, that a living animal requires the presence of an unobservable entity, "a ghost in the machine." To say that a human is "nothing but" a collection of parts is to miss what is distinctly human—the way the parts are organized. All the parts, properly organized, so that they function as an integrated whole, are a human being: "what's what." The "proper organization" of the parts is passed on through gradual evolution from generation to generation through genetic machinery coded in DNA, and by mechanisms we have just begun to investigate.

THE RECOVERY OF NERVE

The foregoing brief chronology is intended to suggest the radical transformation that our conception of life has undergone in the centuries since the Enlightenment. In 1800 there was little or no new factual basis available, since the times of Aristotle and Lucretius, for anything other than an imaginative and speculative interpretation of life. Today our culture is increasingly permeated with evidence upon which to base naturalistic interpretations of what were once mysterious biological phenomena. People with some exposure to that evidence are thereby increasingly enabled to connect the conceptual dots themselves.

The two principal areas that have been rapidly opening up to biological research into the "miraculous," in our time at least, are (1) genomics/proteomics, the science of gene expression in the form of different proteins (the workhorses of life) under different conditions, especially in the developing embryo, and (2) the extraordinarily complex science of brain function. The knowledge that is capable of being discovered in both these fields is almost unimaginably powerful. While recent advances in both of them are frequently described in exaggerated language as "spectacular," the truth is that we have barely opened the door to the improvements in human life that can be made possible by continued investigations in these fields. For instance embryonic development, a perennial source of awe and wonder, is destined for early demystification, with far more control over the course and outcomes of development and also a far better understanding of its mechanisms.

In any case, the mechanistic view of life seems to be gaining ground steadily with the apparently inexorable advance of scientific knowledge. To the extent that scientific inquiry confirms some mechanistic hypotheses and disconfirms others, as it is designed to do, it is seen as a successful enterprise, by its adversaries as well as its advocates. As more and more opinion-makers accept the mechanistic view, with its implications for human nature, we may eventually expect certain consequences to flow from this view for mainstream bioethics. While we will continue to assert a human machine's responsibility for its actions, and its intrinsic worth and dignity, at the same time we will recognize that these concepts can no longer be based on traditional dualisms.

Though the mechanisms of ontogenesis from a fertilized ovum have only just begun to be investigated, we can be certain that in a matter of decades they will be known in minute detail. As they are elucidated one by one, we may envision profound changes in the way most people view the process. It must inevitably dawn on thinking persons throughout our culture that the "miraculous" development of a human embryo into a newborn infant can only be understood as a series of physical processes—of great complexity, to be sure, but without any need for supernatural intervention. Well before that day arrives, most of today's restrictions on embryo research will have been removed, and the concept of an emergent "soul" in the embryo will have become outmoded. The notion of the worth and dignity of a fertilized egg will have been replaced by the more rational concept of gradual organization of the embryo into a machine of enormous complexity and beauty. Notions of absolute prohibition on abortion will have been discarded in favor of morally nuanced issues such as "at what stage in its pre- or post-natal development should a human machine acquire moral status and the protection of the law?" And "what modifications, if any, in the human machine's genetic instructions should be permissible, and why, when, etc.?"

With new discoveries in these two fields, new paradigms, new modes of thought are being developed for dealing with these subjects in clear and realistic fashion. This is a style eminently suitable for practical uses, as contrasted with the impressionistic, even sentimental, but largely passive and ineffectual way our culture has usually treated these subjects up to now. One example of this romanticism is the fascination we have with photographs of the miraculous thumb-sucking embryo. If we Humanists are to lead, we must move into the forefront of our society in efforts to assimilate and disseminate fundamental concepts and scientific data in the pertinent fields of research. We must accept responsibility for a close examination of the risks and the benefits likely to be associated with various proposed applications of the new knowledge. Admittedly, even at current rates of progress it is already becoming very difficult to attain, let alone maintain, the expertise required to achieve this goal, but we must try.

In his *Five Stages of Greek Religion* classicist Gilbert Murray described a widespread loss of confidence in the ability of science and reason to solve human problems during the Hellenistic period of Mediterranean civilization, and a concomitant slide into supernaturalism and withdrawal from the world. He used the phrase "failure of nerve" to account for this phenomenon, and many later authors have used it similarly. I have borrowed the phrase for the title of this essay, not merely for the play on words that it suggests, but also because I believe the best hope of our society for the future of the world lies in a "recovery of nerve." By this I mean the cultivation of a cautious optimism about our collective ability to use powerful knowledge and technologies wisely. A number of truly awesome technologies have already been unleashed in our midst and many people fear that these, or others even more powerful, may already have become unmanageable. It would appear that we have no alternative to harnessing them, and the sooner we set about mastering the skills for doing so the better it will be for all of us. Moreover, the very technologies that are so frightening to us may eventually turn out to be essential for our salvation.

PITFALLS IN BIORESEARCH AND BIOTECHNOLOGY

Some of the dangers of biological research and technology are very real and very troubling. It would be a disastrous mistake if, in our enthusiasm for their benefits, we were to overlook potential problems or make light of them. In fact, the benefits can themselves become very serious problems when they are distributed unevenly and unfairly. To be sure, this is often difficult to prevent, given the nature of the demand for these benefits and the physical and educational resources and huge capital investments that are typically required to produce them.

Biological research can be very expensive and very slow and frustrating. It may require the efforts of large numbers of people to obtain sometimes marginal outcomes. Parenthetically, the greatest numbers may be required when the outcomes are the most marginal statistically. When the results appear to warrant technological developments, further investments of capital and highly trained personnel are usually necessary, often with great uncertainty as to the prospects of success.

To recoup their investments producers often resort to onerous patents, high prices, and outlays for expensive advertising, for which the consumer ultimately pays. Conventional market forces simply do not work in those cases where the demand by part of the market is insatiable. New industries spring up overnight as a result of advertising hyperbole and the wishful thinking of highly vulnerable and poorly educated consumers, concerned about all kinds of "health" problems from acne to death. Princeton medical economist Uwe Reinhardt points to sharp cultural and national differences in patient behavior with his comment that "death is not an option for Americans." The result in the United States is that overall medical expenditures as a percentage of gross domestic product continue to rise with no sign of tapering off, while the percentage of the population unable to afford basic medical care continues to increase.

FREEDOM—AN EPILOGUE

At the opening of this essay I pointed out that in the very broadest sense of the term all ethics is bioethics. All ethics deals with how the behavior of one living organism affects another living organism. Certainly the question of the meaning of "choice" on the part of a living organism lies at the root of all bioethics. If a human organism is a special kind of machine, then in what sense does a human have "free will," often defined as the ability to act in a way that is not determined by causal physical laws? It is surely normal for us to feel that "we" are in control of our own actions, and indeed it would be decidedly abnormal if we did not feel this way. But, apart from our feelings, society's laws and the rules of ethics hold us to be responsible for our actions. How is this possible, if we are simply acting according to the dictates of our physical neural mechanisms and have no power to transcend them?

If the universe obeyed Newtonian mechanics, there would be no possibility, not even a theoretical one, of "free will," interpreted as an evasion of neural causality. In response, the advent of the "uncertainty principle," has been seized upon as an opening for some sort of free will by those who wished to believe in it. After all, if there is some uncertainty in the machine there is some "wiggle room" in which free will might operate. There are, however, at least four major objections to this notion. First, it seems to depend on some form of dualism, involving an immaterial mental substrate ("ghost in the machine") that can interact with physical matter. The idea has been likened to "molecular telekinesis." Second, the randomness implied by the "uncertainty principle" leads to an unsatisfactory kind of linkage between will and action. Whatever else it may be, "free will" is not generally considered a crapshoot. In the third place, it is not clear that the "uncertainty principle" has significant effects on neural functions, which are thought to be highly redundant. And finally, even if it has measurable effects there is as yet no evidence that something like "free will" actually occurs. In fact, such experiments as have been designed to measure the temporal relationship between a conscious decision and the "resulting" *voluntary* act appear to show that the act actually *precedes* the conscious decision to act, thus making the "agent" merely an observer of the action. Ingenious models such as the one devised by physicist Sir Roger Penrose in an effort to circumvent these and other objections have not yet convinced most neuroscientists. Insofar as we perceive that our actions conform to our wills, it might be said that we are "free" to act as we will to act, but our will itself is not "free" from the laws of physics.

When the preliminary sequencing of human DNA was completed a few years ago, the media were fascinated by the possible implications of "genetic determinism" or, as it was sometimes called, "biological determinism." At a seminar on DNA conducted for a group of awestruck judges, some said things like,[4] "The entire basis of Western jurisprudence will have to be rewritten now," and "This undercuts our whole system of laws." Commentators pointed out that social scientists had long been aware of "cultural determinism" and asked whether this should be taken into account as well. The reaction would surely have been even stronger if the judges and the commentators were given a similar seminar on the strong causality implied by the concept of "the nervous system's machinery." It would presumably include not only the discrete influences of genetic and cultural determinants but also the moment-to-moment details of billions of cause-effect relationships underlying our behavior.

A number of accounts of "free will" from the viewpoint of neuroscience have been given by a variety of scientists. One of these is by Patricia Churchland, who coined the term "neurophilosophy" and took it as the title of a book published 14 years ago. In Chapter 5 of her new book *Brain-Wise: Studies in Neurophilosophy*,[5] Churchland endorses the lucid exposition of David Hume in *A Treatise of Human Nature* (1739), where he explains that we do not consider an agent's choices to be "freely" made unless they are *caused* by his desires, intentions, and beliefs: "where [actions] proceed not from some cause in the characters and . . . disposition of the person, who perform'd them, they infix not themselves upon him, and can neither redound to his honor, if good, nor infamy, if evil." If his actions do not have such causes, Churchland adds, "his sanity and hence his control are seriously in doubt." From this analysis, Hume concluded that in ordinary practice we use "free will" to refer to caused, not uncaused behavior, and Churchland concurs. She adds that "the brain does indeed appear to be a causal machine. So far, there is no evidence at all that some neuronal events happen without any cause. True enough, neuroscience is still in its early stages, and we cannot absolutely rule out the possibility that evidence will be forthcoming at some later stage. Given the data, however, the odds are against it."

Obviously these issues have been examined many times in the past.[6] Equally obvious are the semantic problems evoked by the established connotations of the word "machine." Neither of these should prevent us from raising such issues again. For a Humanist the answers to the basic ethical questions may be self-evident. We are all to *be held* responsible for our actions, and therefore we must all *accept* responsibility for them. Our survival as humans living in complex communities depends on the principle of *homeostasis*, by which organisms and communities detect important "errors" (deviations from an internal steady state) and correct them so far as possible.

Some sense of "free will" seems to be a universal experience. If the concept of "free will" is helpful or perhaps even necessary for survival, it will be retained as a social convention, a useful fiction if nothing else. The "worth and dignity" of every human machine still makes good sense as what it has always been, an ethical axiom.

In my view the mechanistic view of humans can level people's expectations of each other, help them to become less self-righteous and more realistically neutral in evaluating social relationships. In the area of law, the mechanistic view helps to reduce emphasis on revenge and vindictively harsh punishment and substitute a greater emphasis on prevention and correction. Legal minds will ponder what sort of re-educational or security system would be appropriate for protecting society from wayward human machines.

Any new technologies should always be applied with great caution, because of the ever-present possibility of unanticipated consequences. But the oft-repeated mantra that we should never tamper with anything so "perfect" as a human being is itself a dangerous canard, as was noted by Charles Darwin in a quotation cited in the title of Richard Dawkins' most recent book, *A Devil's Chaplain*: "What a book a devil's chaplain might write on the clumsy, wasteful, blundering low and horridly cruel works of nature!"

As for queries about specific social issues of bioethics—abortion, embryonic stem cells, genetic modification, and so on—the demystification of living organisms and, in particular, the view of humans as machines should help steer the discussion away from ideas resting on traditional dualisms and toward answers more amenable to reasonable analysis and the determination of facts.

At the very least, I believe that "free" and open consideration of the true meaning of "freedom" will add realism to the internal and external dialogues of Humanism.

Freedom does not consist in the dream of independence from natural laws, but in the knowledge of these laws, and in the possibility this gives of making them work towards definite ends.
—Friedrich Engels, *Anti-Dühring* (pt.I,ch11)[9]

NOTES

1. Information from the Internet.
2. Cited by Schafer, David, "A Scientist Looks at Evolution and Religious Humanism," *Religious Humanism* (3/4), 1996, pp. 12-13.
3. For definitions of types of reductionism see Honderich, Ted, *The Oxford Companion to Philosophy*, Oxford University Press, Oxford, 1995, p. 750.

4. Feigl, Herbert, "Logical Empiricism," pp. 3-26, in *Readings in Philosophical Analysis*, ed. Feigl, Herbert, and Sellars, Wilfrid, Appleton-Century-Crofts, New York, 1949.

5. Murray, Gilbert, *Five Stages of Greek Religion*, New York/Oxford, Clarendon Press, 1925.

6. Taken from a news report in the *New York Times*.

7. Churchland, Patricia, *Brain-Wise: Studies in Neurophilosophy*, Cambridge, MIT Press, 2002. For the citation from Hume, see p. 411 of Hume, David, *A Treatise of Human Nature* (1739-40) Vol. II, Part III, Sect. II, ed. L. A. Selby-Bigge (1888), 2nd ed. Rev., P.H. Niddich, Oxford University Press.

8. See for example, University of California Associates, "Freedom of the Will," pp. 594-615, in *Readings in Philosophical Analysis*, ed. Feigl, Herbert and Sellars, Wilfrid, Appleton-Century-Crofts, New York, 1949.

9. Marx, Karl, and Engels, Friedrich, *Basic Writings on Politics and Philosophy*, ed. Feuer, Lewis S., New York, Anchor Books/Doubleday, 1989, pp. 278-80.

3.

Bioethics of the Germline

Andreas S. Rosenberg

Ethics is a nice Greek word for the list of acts that groups of humans consider and have considered as unacceptable for whatever reason. Historically, academic efforts have tried to define this list-making as a science or more precisely as a science of establishing norms (a normative science). This, of course, is a more impressive way of describing the process of list making. Efforts to deduce the content of such lists of acts from some simple general principle, such as Kant's "categorical imperative," have been attempted again and again, but quite unsuccessfully. Any act has so many consequences in so many different circumstances that necessary clarification and definition of all relevant variables make simple commands, such as **Do!** or **Do not Do!** quite illusory. How can you for example talk about sanctity of life if you torture and kill animals for the pleasure of eating? Aha! We make a shift of frame and state that because animals have no souls, we may eat them. So, who has no soul? Those we eat. What an ingenious piece of logic. But, I do not eat sea anemones. Aha! Maybe this proves they have a soul.

If we define soul as the property of humans, we at least can agree that we certainly may not kill humans. However, if we put on military uniforms and ask our chaplains to bless our weapons, killing humans becomes permissible, even laudable. We have to shift the frame of reference once more. There are evil humans and, therefore, evil souls, i.e. those you can kill. What then about killing an innocent bystander when your bomb goes off course? Well, this is only "collateral damage." Notice that another shift of frame has taken place. And so we go, on and on. The point I am trying to make is that there is no principle from which our ethical rules can be derived. They are just lists of acts that we at this point find useful to declare unacceptable, immoral.

For practical purposes, the most effective way to explain the rules has been to declare them as being ordained by Gods. The arbitrariness and inconsistency of ethical rules can then easily be explained by the jocular and ironic nature of Gods who enjoy playing games with us. As the biosphere evolves and human tribes develop, the rules of ethics seem to change. But, they remain focused on promoting tribal coherence, group spirit, and the illusion of achieving the protection of the Gods. Concepts such as good and evil, just and unjust are human tools, different in function but no different in kind from hammer and saw. A functioning ethics is a useful tool for sustaining human society. In strictly localized circumstances—a family, a tribe, even a city-state— we can talk about a science of making rules for reaching a goal. If we can define what we want to accomplish, we can develop effective rules. To achieve smooth high-density traffic on roads we developed traffic rules. If, however, we try to apply this to human behavior in general, we run into serious problems. Generalized goals are either too various or undefinable or empty of content. The goals of human evolution and social development are not only ill defined but hidden in the fog of future.

Focusing on humans and their history instead of on the Gods does not help much.

In the fateful year of 1859 [1] society learned in a definitive way that humans had no souls. Human beings were just another species of warm-blooded animals with skeleton. Some members of society immediately understood that the observation that big fish ate little fish had immense implications for us. Nature it seemed was suddenly filled with claws and sharp teeth and blood-red was its primary color. Forgetting all about the soul, we can insist that nature itself has defined us as an aggressive carnivore. The only reason that we do not eat each other is that this would diminish our ability to fight other animals. Can a meaningful ethics be built on such premises? No. To the contrary, our nature deviates from the nature of claws and teeth, at least to some extent. Our nature requires civilization and the rules of ethics, the rules we humans have accepted as foundation of civilization.

Yet, civilized though we may think we are, how shall we understand boiling crustaceans alive and eating oysters while still alive? Typically, we move these acts out of sight, hiding them behind the screens of civilization—the stage screens of civilization that are designed to create a false picture of reality. Perhaps, the aborigines follow the brutal laws of nature but we, the civilized people, stretch our little finger daintily out from the tea cup and expect our hired hands to do the killing behind the screens. As for the ever-present blood sport of killing each other, we have made it palatable by ritualizing it. Colorful uniforms, shining helmets, glittering medals, martial music and television coverage makes war almost a sporting event—a quite civilized behavior with some regrettable casualties. For these, there is always of course the proper ritual of washing ones hands or saying a prayer.

This was the ethical setting for our society and for our individual lives. For the western world, it was a successful arrangement, allowing us to preach morality to the pagan while enjoying the hidden benefits of deceit. Once in a while, however, storms unavoidably shake our civilized teacups. Circumstances change and the convenient scenery of civilization has to be changed to fit the circumstances. What was right becomes wrong. One such storm was the struggle over the abolition of slavery, a practice that had supported civilizations for thousands of years. During all that time, it was ethical to buy and sell humans, to treat them as property, to keep them in chains, to make them row our galleys and harvest our fields. If they had a soul, it was definitely a second rate soul. Consequently, most churches, like most people, supported slavery.

However, this comfortable pyramidal model of human society with different values for different levels of the population was challenged by the thinkers of the Enlightenment. They brought forward the new concept of equality which was established in the blood of the American and French Revolutions and the revolutions which followed it. Equality and freedom became new basic moral rules for a new political model, democracy.[2] The economic advantages of slavery, however, allowed the practice to linger on until the middle of the nineteenth century in more underdeveloped societies such as Imperial Russia and United States. In United States traces of it could still be found until the civil rights crisis of the 1960s.

We are today again in a stormy period involving an aspect of ethics, a sub-division aptly called bioethics, the ethics and morals of biology. Biology in this context is the manipulation of living material. It includes medicine, agriculture, zoology, botany and related fields. The ethical problem, of course, arose when we started to manipulate human material. For years we had the attitude that as far as animals and plants were concerned, anything goes. The old problem of the unique position of humans as the only ones in possession of a soul appeared again. As far as humans were concerned, the rules had been simple: For example, we agreed that all newborns had a right to survive. We could, of course, kill each other but only under complicated ritualistic rules. Similarly, we agreed to let the nature take its course and not to shorten the natural life span of aging humans with the possible exception of assisted suicide. If we add to these the rule against humans eating humans, we have bioethics to date, complete and functional.

Governing the actions of physicians, the only ones allowed to manipulate human life, was not really a problem at all. It was really an issue of law and licenses, and in some senses no different from setting up rules for car mechanics. Physicians were responsible for doing their best in their direct relationship with their patients. By and large, they were not primarily responsible for increases in life expectancy and improvements in quality of life. These were instead due to public health efforts, to changes in sanitation and nutrition. The question of prolonging life by extraordinary measures was solved by legalizing the right of the patient to decide on withholding or withdrawing treatment. Abortion as well as family planning was seen as a sociological or religious problem and not a medical problem. In dealing with these and most questions of bioethics, the role of physicians as physicians has, contrary to popular press, been marginal.

So far so good. Life, as we know, follows a torturous causal pathways. The results are unpredictable in detail. Individual happenings and choices can be perceived as random, coin tossing type of events. But, they do form like pearls on a string, an ordered sequence of happenings in time. We use the concept "stochastic process"[3] to describe such chains of pearls. *The word* stochastic *also implies that, although the exact outcome of a sequence of events cannot be predicted, the possible outcomes form a well defined array of choices.* For example, we cannot predict anybody's life expectancy at birth, but we know it falls in a distribution of life-times, with an average defined by the most probable life expectancy for the country we are dealing with.

Additional choices in one's life are primarily economic and social. These also are of stochastic nature. So, the primitive ethics I have described took adequate care of the few choices most humans had to make on the biological level. Life is really a game of chance starting with birth and the choices of parents through a complicated path through wars and economic and personal crisis to an undefined but surely final endpoint.

> *To summarize thus far: from a biological point of view, the choices for any human being are few and with unpredictable results. So, personal choices on that level are nearly non-existent. If you have no choice, you have little use for a more complicated bioethics than the simple rule of not killing and not eating your neighbor.*

Whereas bioethical choice may not be terribly relevant for a single human being from his individual point of view, however, it has grave implications for him as a member of the human race. It is here that the title "Bioethics and the Germline" becomes important.

Manipulation of living organisms can take place on two levels. 1) The **Somatic** level, which deals with changes or repair of the majority of our cells, and which do not influence our heredity, our descendants. 2) The **Germline** level, which deals with changes in the DNA code that is present in our ovum or sperm. These are passed on to our descendents. Somatic changes die with us while germline changes become a part of the future of the race. Any human-induced germline change becomes part of the natural stochastic process we call evolution.

The major contributor to the stochastic nature of our trajectories in life is the chromosome shuffle. The 23 chromosomes from the mother and the 23 from the father are combined in an unpredictable manner to provide us with 23 pairs of our own chromosomes. The presence or absence of most inherited traits is again a stochastic process. My mother had six sisters and my father the same number of brothers. How they all behaved and looked was defined by both the genetic components from the chromosome shuffle and the structure of the memes—patterns of behavior and understanding acquired while growing up and stored in our memories. All the siblings turned out to be very different; some similarity could be found only in their facial structures. With some difficulty, you could identify them on old photographs, both the unexpected successes and unexplained failures. From two parents came twelve vastly different individuals.

The chromosome shuffle is the foundation of the process of evolution.[4] It provides nature with the means to try out every possible variation of life-forms, sample every possible niche in the biosphere. This is how we humans raised ourselves on our hind legs and became thinking apes with the ability to talk. The process of variation of possible outcomes does not stop, however, with the formation of the embryo. The differentiation of our cellular components and the formation of an organism is also a stochastic process. Thus, further random variation is introduced. The outcomes are unpredictable and surprising. This is how we got Mozart, Luther, and Stalin.

What has happened recently and what is so revolutionary is that by uncanny energy and cleverness we have developed an understanding of the mechanism of the chromosome shuffle. We know the rules of the game, we have the pieces used in it, the genes, and now, finally, we have the ability to draw the map of the game, the human genome (the sequence of all the genes!). The sequence of every one of the 23 pairs of chromosomes can be determined. With the map of the pieces and a book of rules, we are ready to play the game. Knowing the game, we can manipulate our genome intentionally. We can bypass the chromosomal game of dice and eliminate, at least partially, the stochastic nature of the outcomes. Of course, in a more primitive fashion, we've done it before. By careful mating, a controlled shuffle of the chromosomes, we, for example, developed the race of mini-poodles whom we then taught to roll over on command. It is important to realize that dogs roll over on command and humans speak English because a system of memes has been created in each of them for each purpose. We have been well on our way to cloning, gene therapy, and genetic counseling even if in the beginning only in lower animals. These poodles we believed were "soulless" and thus whatever we did was not considered unethical. Now, however, we are on the verge of cloning the human being, beings with a "soul," and the problem of clones, gene therapy, gene modification have became the objects of intense discussion. This discussion can be reduced and combined into one single question.

> *Is our ability to change and manipulate the human genome a cata-strophic and deadly insult to the normal, natural mechanism of evolution and thus an imminent danger to human race; or is this recently acquired ability a step in the normal process of evolution, a step forward for homo sapiens?*

All putative steps in evolution have to be tested and so does this one. What will be tested is the viability of self-regulated evolution. We can speed up human evolution; we can increase our memory and eliminate the potbellies that are threatening our culture. Is this desirable or not? Our problem at hand is that choosing between the two alternatives outlined above is difficult. We have to project into the far future, over many generations—a process we are not accustomed to. We have to produce scenarios for future development.[5] To do that, we have to reduce the complexity in real life to some variables which we arbitrarily think will play a major or decisive role in the future.

IN WORKING ON THIS QUESTION, I HAVE CHOSEN TO FOCUS ON TWO COMPETING FORCES

The first is the gross global product of resource utilization and waste production, a process that rolls ahead without any rational plan. I include in it population growth, pollution and destruction through conflict.

The other force is a growth in the human ability to reason and its spread through education.

What we have to assess is the speed with which these two forces increase. We also assume that the first, driven by population growth, technology, religion and individual greed, is ruled by short term goals. Considering the time scale of Darwinian evolution, the challenges produced by any meme-driven changes in society are too fast for a response by the classical Darwinian mechanism. In the Darwinian mechanism a successful change is marked by a change in some manner in the DNA when compared to the unchanged condition of the rest of the genome. This just takes too many generations. The second force, that of increase in ability to think rationally, is based on a premise that enlightened people recognize danger and act for self-preservation, i.e. they act in a non-random fashion. One example is Northern Europe where the political and social discourse has been quite rational, leading to rational regulation of human endeavors. Although the consequent policies themselves are also meme-driven and thus too fast for the Darwinian mechanism, the basis for memes—residing in the human brain—is indirectly malleable by Darwinian mechanisms. Should changes towards this kind of behavior be helped along by intentional evolution focused on the human brain?

What are the major forces that drive the curve of gross consumption? The Malthusian nightmare of the human population increase is still with us.[6] We are six billion today and increasing. The Malthusian problem is often denied these days by those who, for example, consider the green revolution an adequate response to it. We will be able to feed any population; there is no direct sustainability limit. On the other hand, there is no answer yet to the other side of the population growth argument, that of the disposal of waste. The stream of waste and the stream of chemicals necessary to sustain increased agricultural production will by all calculations slowly but steadily poison the biosphere.

It has also been pointed out that the high level of technology and the nearly certain high life-expectancy lead to a voluntary decrease in the rate of reproduction, a phenomenon already visible in Northern-Europe. However, if the population of the world we have to-day is to be brought up to the living standards of Northern Europe, the gross consumption curve would skyrocket. An increase of population from 6 billion to 18 is not different in its demand on the planet from the alternative of increasing the consumption of the present six billion threefold.

The rate of consumption, the stream feeding the gross national product (GNP) of any country, is a sum of the number of individuals and the average consumption and waste production per individual. I am sure we all have speculated about what will happen if one and a quarter billion Chinese and one billion Indians reach the level of one automobile per three persons! The process towards this number is already underway. I have just returned from Vietnam where the population, now 80 million, has already acquired 3-5 million motorcycles. With increasing wealth, the conversion to automobiles has begun. The beautiful beaches of Vietnam are already covered with refuse, thicker and thicker layers of plastic bags. Of course, there will be forces working in our favor like education. There will be changes in technology shifting behavior toward recycling and permitting a more efficient disposal of waste. But these are slow and difficult social processes heavily dependent on education.

At every place there are two competing streams of happenings: first, mass consumption driven by technology and the profit motive and second, education and the development of rational thought processes. Which of these has a higher rate of increase? When we look around, the answer is clear. Education lags, hidden behind a wall of advertisements touting that increased consumption as good for you, the nation, and the world. A one percent diminished sales figure for retail is considered economically dangerous. The time-honored name of *citizen* has been changed to *consumer*.[7]

In addition there are malevolent forces working against any limitation of the ever-increasing population shopping incessantly in ever-increasing number of shopping malls. Profit is linked directly to consumption. More disposable plastic bags in Vietnam means bigger profits for some international concern. The intentional linking of the state of a nation to its GNP encourages the production of hula-hoops, widgets, and bigger and bigger automobiles. To these economic and political forces are added religious ones like the Roman Catholic church, still stuck in its medieval mind set. All these variables can be studied separately. It soon becomes clear that the complexity of the development of our global society is such that direct predictions are not possible. What is possible is to develop extreme scenarios as guidelines. The overpopulated consumer world slowly poisoning the biosphere is one such very probable scenario.

Another decisive factor besides the rise of gross consumption and population is the incredible increase in bacterial and viral populations, populations that are directly proportional to the available biomass of human population! We know that the mutation rate of these organisms is faster than ours, despite the fact that the basic rate of mutation per nucleotide pair may be the same. The structure of bacteria and viruses is much simpler than ours. It takes the virus fewer concurrent or consecutive mutations to develop a more virulent strain than for us to acquire the appropriate immunity.

The recent record of our efforts to produce vaccine against just one virus, AIDS, is not very encouraging. A new mutation of AIDS or *Ebola* might well strike us, helped immensely by the rapid increase in air travel. Thus, we may lose the competition with these organisms not because of the impossibility of developing suitable defense mechanisms but because the rate of introduction of new variations of old nemesis outstrips our rate of developing counter-measures. We do not want to accept the fact that we have to devote a major portion of our resources to increase our capacity to keep ahead of likely viral attack enhanced by the effects of population and consumption growth. Instead, we devote these resources to what we miscall *security*, forgetting that no wall can defend us if we are attacked from within. Pericles surrounded Athens and Piraeus with impregnable walls and furnished Athens with a large fleet to bring in all necessary sustenance. However, in these close, isolated quarters, plague struck without the help of bioterrorists, depleting Athens' population, Pericles included. The victory went to the Spartans. No walls had to be breached.

Looking at these time-dependent processes—the increase in gross consumption and the development of new diseases, the slower process of intellectual growth—we must ask whether or not our ability to manipulate the human genome does not provide an evolutionarily useful ability to increase the rate of intellectual development and the progress of rational thought?

What we see in our country is that the rate of technological innovation that drives the rate of consumption has far outstripped the rate of intellectual growth in our population. We might argue that the deficiency, so blatantly demonstrated by the tested abilities of our children, is just all a problem of incorrect memes, incorrect learning processes, or unsuitable surroundings. This may be so, but it also tells us that for a majority of the population, the necessary mental prerequisites for a rapid learning process may not be there. As an example we might consider that the knowledge of more and more complex computer and internet systems is a basic necessity for the future. If that is so, then it is probable that the fraction of people without the ability to understand network logic will grow. They may include more young Mozarts or Walt Whitmans. Not all individuals can be taught the ability to manipulate abstract structures and yet the need is there. The possible self-regulation of genetic change is thus an evolutionary opening to correct that shortfall.

If we make a choice in favor of a beneficial and necessary tinkering with our heredity, what are the consequences for a global bioethics? What we have to do is more complicated than the convenient collection of general **Do Not's** of an earlier time. As citizens, as human beings, we have to look to each procedure, each project, each policy and try to estimate its consequences in the near and far future. We have to ask, do these produce the desired increase in the variability and diversity of the human species or do they work against the desired speed in evolution? We have to agree that the increase in the speed of evolution is, indeed, the goal we must strive toward.

Consider some simple examples. It does not take much reasoning to see that any form of human cloning does not contribute towards the desired goal. In fact, it is essentially counterproductive. A nation of copies might be all wiped out by an accidental virus, carrying a mutation lethal for just that human prototype. Gene-therapy on somatic cells may cure diseases and that is not unimportant. But it has little or no effect on the process of evolution. Thus, the question of using or not using stem cells is from a global bioethics point of view largely irrelevant. It is in reality a religious and/or social problem on the level of either eating or not eating meat on Fridays or driving on the right or left side of the road.

When we propose modifying germline DNA, we have a problem on our hands. If and when we change a gene or introduce a gene, the change will be carried by means of the chromosome shuffle into the human genome, become part of the human genome. In this way, we might increase human memory, speed up our computing ability, and allow it to store more and more of the complex algorithms that form the basis of rational thought. However, all such manipulation are as dangerous as they are promising. For example, a modification of genes to allow humans a 300-year life span, if it stretches the time required to reach childbearing age, has serious evolutionary consequences. Since evolution counts time in generations and not in years, we would be slowing the rate of evolution. On the other hand, a mutation shortening the life span could increase the rate of evolution by gene shuffle. But, since the transfer of memes demands time, the total effect would probably be minimal.

I have reasoned here that of the two alternatives the one with rapidly increasing gross consumption outstripping our intellectual development has a greater long term risk for humanity than the trials and errors of gene manipulation. In my view, we have to risk trial and error but it must be combined with rigorous societal control of experimentation all within the framework of a world government. Recalling our problems with the control of nuclear weapons and biochemical warfare, this proposal might well seem utopian. But it is necessary. At the same time, it is clear that as long as national governments exist—democratic or not—genomic manipulation would become another weapon in the futile fight of who controls whom. Internationalization become a key to an effective global bioethics.

So, as I see it, we have now to move toward a new set of **no's** and a few **maybe's** for the rulebook of bioethics. I understand that this is not a conclusion based firmly on some biological sociological calculus. We just do not have such rigorous systems yet. So my reflections on global bioethics represent guess work. They are based on some simple scenarios and perhaps scarcely deserve the name of ethics. But ethics, as we know them to-day, are based on a beautiful mythology which for the problem at hand we cannot use and do not need. Instead we need risk a model, a simplified one no doubt, but one where we can assess the longtime consequences of our actions or failures to act.

NOTES

1. D.D. Dennett. *Darwin's Dangerous Idea*. Simon and Schuster 1995.
2. J.B. Bury. *The Idea of Progress*. Macmillan 1932.
3. B.V. Gnedenko and A.Ya. Khinchin. *An Elementary Introduction to the Theory of Probability*, Dover Publications 1962.
4. Dennet, *op. cit.*
5. T. de Chardin. *The Phenomenon of Man*. Wm Collins 1959.
6. P.M. Hauser ed. *The Population Dilemma*. Prentice Hall 1963.
7. J.K. Galbraith. *The Affluent Society*. Houghton-Mifflin 1959.

4.

The Moral Status
of the Human Embryo
The Twinning Argument

Berit Brogaard

Recent scientific advances in research involving stem cells derived from human embryos have sparked considerable ethical debate concerning the moral status of the human embryo. Scientists believe that research using stem cells might eventually help us cure diseases such as Alzheimer's, Parkinson's, juvenile diabetes, spinal cord injury, and diseases of main organs. Stem cells apparently have the capacity to transform themselves into specific organ tissues. In the future, researchers may be able to use stem cells to develop, for example, liver cells that could cure someone with a malfunctioning liver. Because human embryonic stem cells have the potential to develop into all of the tissues in the human body, stem cells from human embryos are believed to have a greater potential to produce these results than stem cells from adult cells, umbilical cords, or the human placenta.

For research, embryonic stem cells are isolated from human blastocysts—embryos at around day four in the fetal development. In the course of obtaining stem cells from a living human blastocyst, the blastocyst is destroyed. For that reason, the main ethical challenge associated with stem cell research has to do with whether blastocysts have any moral status. There are two main arguments for the view that blastocysts have moral status. One relies on the claim that, if the embryo under favorable circumstances would be identical to the human being as it would exist after birth, then it already has the same moral status as a human being after birth. The other relies on the claim that the embryo is entitled to protection because from the moment of conception it is a potential human being, even if it would not be identical to the future human being under favorable circumstances. The second of these arguments will not concern me here.[1] My focus is on the argument that the embryo has moral status because under favorable circumstances it would be identical to the human being that would exist after birth. I shall argue that the premise is false.

The most important argument against this claim, I believe, is the twinning argument.[2] The twinning argument, in its simplest form, is familiar: prior to gastrulation (the critical developmental step at which the embryo's three major germ layers, or tissue type divisions, form), the embryo is susceptible to twinning. Accordingly, it has the potential to develop into several human beings. Since the process of becoming a human being has not yet ended, the pre-gastrular embryo is not a human being.[3] Unfortunately, the twinning argument in its simplest form is flawed. The premise that if an entity is potentially two, it cannot be one is false. As the American Civil War teaches us, there are cases where identity is inherited even though an entity is susceptible to twinning. The United States in the period immediately prior to the Civil War was actually one but potentially two.[4] But, I will now argue, even though there are clear cases of individuals susceptible to twinning that persist over time, there are empirical grounds for denying that an embryo susceptible to twinning can be transtemporally identical to a future human being.

To make a case for the proposal that an embryo susceptible to twinning cannot be identical to a future human being, I shall first consider the possible scenarios under which twinning might occur. Second, for each possible scenario I shall determine whether there are empirical grounds for thinking that human embryonic twinning occurs under that scenario. Finally, for each possible scenario I shall determine whether it is conceivable that the pre-gastrular embryo under favorable circumstances is identical to the human being as it exists after birth. As we will see, according to the scenario under which embryonic twinning is known to occur, it is impossible that the pre-gastrular embryo is identical to a future human being.

HOW SHALL WE UNDERSTAND TWINNING?

First, one may hold that twinning is a form of cloning. On this view, the embryo would, in some pre-gastrular phase, be such that a process of forming new human individuals can still occur via budding. When budding occurs, a part of one individual substance becomes detached and forms a new individual substance in its own right, while the original substance continues to exist. Similar scenarios are known from the vegetable kingdom, where a cutting from one plant may be planted in the soil to result in a new plant without the original plant ceasing to exist as a separate individual. Budding does not entail the destruction of the entity that buds off the new individual. So, if embryonic twinning occurs via budding, the view that the pre-gastrular embryo under favorable circumstances would be transtemporally identical to the future human being is credible. Unfortunately, however, human development is nothing like that of plants. When a cutting is taken from a fully developed plant we have an original organism and a part that develops into a separate individual. We do not have one cell, or mass of cells, which divides into two. But this is exactly what we have in the case of embryos that undergo twinning.

Second, in light of the above considerations, one may hold that twinning occurs via fission.[5] When an entity (for example, an amoeba) undergoes fission, the entity reduplicates itself. New parts are formed that then split apart to lead separate existences. When fission occurs, the entity that undergoes fission is destroyed. This claim rests on the conceptual thesis that one individual cannot be identical to two individuals. Think of a log that is split into two for burning in the fireplace. One may find it quite natural to think that one log is identical with two logs given that their time frames are different. However, identity over time, like all other sorts of identity, is transitive. That is to say, if John-as-a-baby is identical to John-as-a-child, and John-as-a-child is identical to John-as-an-adult, then John-as-a-baby is identical to John-as-an-adult. Identity over time is also symmetric: if John-as-a-child is identical to John-as-an-adult, then John-as-an-adult is identical to John-as-a-child. Compare identity to love which is neither transitive nor symmetric. From the fact that John loves Mary, and Mary loves Peter, it does not follow that John loves Peter. And from the fact that John loves Mary, it does not follow that Mary loves John. Since identity is both transitive and symmetric, if the one log were identical to the two logs, the two logs would be identical to each other. But they are not. We just said they were two, not one.

Of course, one may say that the one log is identical to the two logs taken together. But if this is so, then spatial connectedness does not matter in defining what a log is. That is, something can be a log even if its parts are spatially separated. In the case of an embryo spatial separation may not matter either. But the two embryos that would result from the splitting of a single embryo would under normal circumstances develop into two human beings. So if the single embryo were identical to the two embryos taken together, then the single embryo would be identical to two adult human beings, which cannot be the case. Alternatively, one but not the other of the twins would be identical to the ancestor entity. But there is no property in virtue of which one but not the other twin could be said to be identical to the ancestor entity. So, if human embryonic twinning occurs via fission, an embryo that results from a twinning process begins to exist when the twinning process is completed.

But the fact that an embryo ceases to exist when twinning occurs does not rule out the possibility that an embryo that does not undergo twinning is identical to a future human being. So, one may hold the view that, when twinning occurs, the embryo begins to exist when the twinning process is completed, but, when twinning does not occur, the embryo begins to exist at the time of conception. Since we do not know whether a blastocyst would undergo twinning under favorable circumstances, we do not know whether it would be identical to a future human being under favorable circumstances. And since we do not know whether the blastocyst has moral status, destruction of the blastocyst is not justified.

The aforementioned theory entails that I could not have been a twin if I am not in fact a twin. For if twinning had occurred to me, I would have ceased to exist, and two new human individuals, that are not identical to me, would have come into existence. It may seem counterintuitive that I could not have been a twin. But no objection to the possibility that a pre-gastrular embryo—as twinning fails to occur—is identical to a future human being is to be found here. It is simply a conceptual fact that individuals susceptible to twinning via fission could not have been twins.

As it turns out, however, human embryos do not undergo twinning via fission. The reason is that the cells—and sums of cells—in the pre-gastrular embryo are totipotent: each of them has the potential to develop into a complete human being. The parts of an amoeba or a virus, by contrast, do not have the potential to develop into a new organism. New parts are formed which then split apart to lead separate existences. When the pre-gastrular embryo undergoes twinning, the bits destined to become a human being split into two halves without any foregoing reduplication.[6]

Recent embryonic research has given rise to skepticism about the view that the embryo is a featureless orb of cells prior to gastrulation. It has been shown that the mammalian body plan begins to be laid down already from the moment of conception.[7] A body axis is present, and there is already a clear division between bits destined to become the placenta (etc.) and bits destined to become the future human being, even at the two-cell zygote stage. Despite the existence of traits that appear to narrow down cell fate, the fact is that pre-gastrular embryos are susceptible to twinning. Even though the cells become biased towards producing certain tissues, those biases are not irrevocable. So the new data do not dispute the claim that the cells destined to become a human being are totipotent. After all, the pre-gastrular embryo does not have the kind of structure that would prevent it from separating into parts that can each produce a complete human being.

SEPARATION AS THE WAY: TWINNING OCCURS

This leads us to the third way in which twinning can occur. On this scenario, the embryo is already not one but two (or more) entities, both of which would survive, should twinning occur, to form two independent embryos. That is, the embryo is, in some pre-gastrular phase, such that a process of forming new human individuals can still occur via separation. Separation is distinct from fission in that two or more entities are joined together as one entity, and at some point the relations conjoining the parts of this entity are disrupted in such a way that the previously attached individuals continue as separate new substances. In the case where twinning does not occur, one may hold one of the following views: one of the two parts of the embryo is under favorable circumstances transtemporally identical to the human being that exists after birth. Alternatively, one may hold that both parts of the embryo in favorable circumstances are identical to the human being.

The first of these alternatives implies a peculiar process under which—as twinning fails to occur—one human being would absorb into itself another entity that is of exactly analogous form and structure. But more important, it leaves open the question as to what might make it true that one but not the other half of the total embryo as it exists prior to gastrulation should be the human being that exists after birth.

On the second alternative, an embryo that does not undergo twinning consists of several parts that in favorable circumstances are jointly identical to the future human being. To show that this alternative is not possible either, I need to employ the notion of being identical to a human being whose possibility was never realized. The notion of being identical to all sorts of possible human beings is not strange at all. If I had not left my hometown I would not have gone to Buffalo nor have married Joe. If I had gone to Ohio State University instead of the University at Buffalo, I would have met Joe earlier than I actually did. These conditionals are not true of anyone else but me, and they are true of me. So, I am identical to a possible person that was never actualized, given my decision to leave my hometown and attend the University at Buffalo. Notice, however, that for each possible scenario there is only one possible human being to which I am identical, never two or three. So I am identical to just one possible human being in the possible situation in which I go to Ohio State University instead of the University at Buffalo. That is, I am identical to the possible human being (in that possible situation) with whom I share most of my qualities in common, including the quality of having originated from particular sperm and egg cells.

I can now turn to my argument that the second alternative is not possible either. Call the two halves of the embryo prior to gastrulation when twinning is still possible A and B. Even though twinning does not in fact occur, there is a possible situation in which A and B separate and continue as separate human beings. The actual human being that exists after birth is transtemporally identical to the actual sum of A and B. But the actual sum of A and B is identical to the possible sum of A and B. The possible sum of A and B, on the other hand, is identical to the possible sum of the two possible human beings. By transitivity, it follows that an actual human being is identical to two possible human beings that are both located in one and the same possible situation. But one human being cannot be identical to two possible human beings that are both located in one and the same possible situation. So the second alternative is not possible either.

An opponent may now object that the bundle of cells constituting the embryo is a whole, and that spatial separation of the two halves of a bundle is enough to destroy the whole. So, even for the case of twinning by separation, the loosely connected bundle of cells is a whole that ceases to exist if twinning occurs. But then the possible sum of A and B is not identical to the two possible human beings. So, the aforementioned transitivity of identity is averted.

What counts against this proposal is that the parts of the pre-gastrular embryo destined to become the human being and the parts destined to become the placenta are not spatially separated. If we suppose the degree of spatial separation between two entities x and y determines whether x is identical to a future individual z while y is identical to a future individual v, the parts of the embryo destined to become the human being would in every case fail to be identical to the human being.

Perhaps the objection is rather this: that there are cases where separation happens because the internal relations among certain parts of a whole cease to obtain, resulting in a split along the line where the former internal relations no longer hold. Even if we think of the pre-split thing as consisting of two parts, it is still one whole thing (and not a mere collection of parts). Because of the nature of identity, the original entity cannot be identical to both of the resultant entities.

The problem with this proposal is that the cells in the early embryo form a mere mass, being kept together spatially by an outer membrane. There is no causal interaction between the cells. They are separate bodies, which adhere to each other through their sticky surfaces and which have at this point only the bare capacity for dividing. In view of that, there is no reason to think that the entire bundle of cells, but not any one part of it, has some property in virtue of which it is identical to some potential human being. If each of the parts with a potential to become a complete human being is identical to a potential human being, the inference goes through.

In sum, human embryonic twinning occurs via separation. But if twinning occurs via separation, an embryo susceptible to twinning cannot be transtemporally identical to a future human being. Hence, if blastocysts are worth protecting, it is not because of their prospective identity to entities that we already know are worth protecting.[8]

NOTES

1. Singer and Dawson have discussed this argument in P. Singer, K. Dawson, "IVF Technology and the Argument from Potential," *Philosophy and Public Affairs* 17 (1988:87–104). See also J. Feinberg, "Abortion," in Tom Regan, ed., *Matters of Life and Death: New Introductory Essays in Moral Philosophy*, 2nd ed. (New York: Random House 1980, 1986), p. 267; and D. Hershenov, "The Problem of Potentiality," *Public Affairs Quarterly* 13 (1999): 255–71.

2. In "Sixteen Days" (*The Journal of Medicine and Philosophy*, forthcoming) Barry Smith and I argue that the pre-gastrular human embryo is not a human being. In this note I develop a new version of the twinning argument that is in many ways stronger than the twinning argument presented there. For discussion of the simple version of the twinning argument, see N.M. Ford, *"When Did I Begin?" Conception of the Human Individual in History, Philosophy, and Science* (Cambridge/New York: Cambridge University Press, 1988).

3. L.C. Becker, "Human Being: The Boundaries of the Concept," *Philosophy and Public Affairs* 4, (1975): 340.

4. Most likely, when we say that the United States is potentially two, we mean that a part of the United States could have become detached and formed a new country in its own right while the United States would have continued to exist.

5. W. Quinn, "Abortion: Identity and Loss," *Philosophy and Public Affairs* 13 (1984): 27.

6. Of course, as any horticulturalist knows, many plants can be bisected with the bisected parts developing into complete plants of the same species. Similarly, Planaria are noted for their great ability to regenerate missing body parts. Chop a planarian into two pieces and each piece will develop the missing parts and become a separate planarian; the head of one (or the left side) is identical to the corresponding part of the original; the tail of the other (or the right side) is identical to the tail of the original. If the bisection of plants and flatworms is a form of twinning by fission, then we have to go carefully to distinguish what we are saying about human twinning. For if plants and flatworms, both of which have parts that are totipotent, can undergo twinning by fission, then what reason do we have for thinking that human embryos, which have parts that are totipotent, do not undergo twinning by fission? Granted. However, these cases appear to be cases of twinning by separation (see below). After all, if I can cut a flatworm into two pieces that can both survive, then it would seem that each part of the flatworm that can survive on its own is identical to a potential flatworm. But this means that flatworms and plants the parts of which can continue to exist as separate entities are not individuals in the sense in which human beings are individuals. Rather, they are collections of parts that could survive on their own under the right circumstances.

7. "Your Destiny, From Day One," *Nature Science Update*, 8 July 2002.

8. Thanks to Judy Crane, John Danley, David Hershenov, Richard Hull, Joe Salerno, Ron Sandler, and Rob Ware for helpful comments.

5.

Exploring the Boundaries Between the Scientific and the Spiritual

Harvey Sarles

What the heck is going on in the world of medicine and curing? More visits to "alternative medicine" practitioners than to MD's in these times, looking for . . . what? Where is truth and the best, or at least good advice and treatment? Who are the fakes and con-artists? How does anyone know? What is the truth; where precisely are the boundaries of knowing and the spiritual, pray tell us? Why the current distrust of medicine, the search for cures or a better life-style and life-adjustment? Thus, my question: science and the spiritual?

This essay reflects my personal experience with issues and questions about health and curing. Over the years, I have wandered around the medical profession, was briefly a medical student at SUNY-Buffalo, did research in a medical setting at U. of Pittsburgh for several years, and have had many contacts and study opportunities at the U. of Minnesota Medical and Dental Schools. My intellectual interests in the human body have also influenced my thoughts about health and curing. I meet fairly regularly with cure-givers and other thinkers in a group we have been calling: "The Body Group." For many years I have been recording my thoughts and observations in "The Body Journals."

1. MEDICINE AS A KIND OF FIELDWORK

At the beginning of the summer of the year I would begin (and "finish") Medical School, I watched a television program on the people who visited Lourdes in France, looking for a cure for what *ailed* them. Having visited Lourdes, and done what they were inspired to do, some of them claimed complete recovery. I was puzzled. What is medicine, sickness, curing? What is health? How can belief or faith, cure?

That summer, I attended a couple of tent meetings of evangelical Protestant groups. Here were people attending, and giving testimony to some "fact" of their recoveries. "The Lord is in the tent tonight! Those who need help, come forward and truly receive the Lord and you will be cured." I remember well the beatific looks on their faces. Hmm.

So what is sickness, what is a cure, and the next question struck me: what is health or healthy? My one year and a summer of medical school didn't much address this issue. Sickness or pathology was the dominant issue in that setting. Patients presented themselves (as often, their families did), and the problem was what to do to cure, a reasonable set of procedures, medicines, etc.—the medical trinity: diagnose, treat, cure.

I also worked in a *Chronic Disease Institute* that year, and got to know well that at least some *parts* of medicine aim to maintain rather than treat or cure. For many people with neurological or other life defining issues cure is not possible. I noted, as well, that some parts of medical practice—obstetrics comes to mind—were less about sickness and cure, and more about guidance.

I noted, as well, that there were physicians and others who assisted with or took on various tasks which sometimes had to do with curing, and other times with maintenance of health. I think of nurses, aides, physical and occupational therapists. There weren't many midwives during that era when most women during delivery were given sedatives, and essentially "forbidden" to breast-feed their infants.

I attempted to raise my questions in medical school, but they were not well received (an understatement), and I left the field of medicine. But the ideas have remained strongly on my mind ever since.

My interests in the nature of medicine simmered during my

training in anthropology, but rose a good bit during a two-year field-work stint in Southern Mexico. We had with us our infant daughter who had been born in Chicago delivered by that particular physician who had written "the book" on natural childbirth (Sol de Lee). So I wondered about fashions and fads in medicine. Now in the Mayan Highlands we were daily made aware of the various medical issues that arose in a very different cultural setting where people had very different ideas of medicine and curing than I had thought ordinary as a result of my upbringing.

In Southern Mexico (Chiapas), the people distinguished two sorts of illnesses. There were those which could be treated by "curers" (there were two cultures there: ladino-Mexican and Mayan who had some different ideas and practices of curing) and there were those who went to a physician. Dentistry basically meant an extraction. I had three, I think, in two years. In that setting, as well, many kids died of dysentery before the age of two. So I witnessed this dire sickness too. And there were other things which caught my attention, including the fact that many poor people seemed to be chronically ill, almost a kind of fate. Of course, there was no health insurance. Much to think about!

My family survived the two years of formal fieldwork fairly intact and returned to Pittsburgh. My first job was at the Western Psychiatric Institute as a research anthropologist-linguist examining the interactions between physicians and patients in the psychiatric interview. I had ample opportunity to study my colleagues, mostly MD's who had completed their psychiatric residencies, and were working toward becoming psychoanalysts.

During this time, I was offered a full-tuition training grant in order to become a psychoanalyst myself. But I declined on the grounds that my anthropologist-self would lose its observational objectivity in the process of becoming a curer—at least that's the story I tell myself.

Beyond my research, I had ample opportunity to observe the medical world. It centered on psychiatry naturally, but extended to many other fields that surrounded psychiatry from surgery, to internal medicine, to the ancillary professions which supported medicine.

My closest colleague, Otto von Mering—also an anthropologist—was a researcher studying medical values. We talked incessantly about what all this was about. Ray Birdwhistell, my mentor during this period, visited monthly (periodically, we jested) from Philadelphia. He had studied interactions in all contexts, especially psychiatry but had also thought much about the nature of curing having done his own fieldwork among Native Americans and in other American settings. Among the questions: how parents diagnose their children, when a symptom is treated at home, when they go to a physician. Everyone's mother is the first line of consultation, and much about medicine is framed in that relationship.

I also got to observe inpatients from the locked psychiatric ward. Interestingly, I got to see many of them, and usually some members of their families, as they were about to be admitted, since my office happened to be on the admissions floor.

I continued to spend a great deal of time in medical settings at Minnesota, often with physicians who studied with me in courses in Anthropology. Given my interest in the body, some of my students who became physical anthropologists worked in various medical setting ranging from anatomy to surgery to physical medicine. My next-door neighbor was a pathologist; others were surgeons. I attended seminars in Oral Medicine/Dentistry for several years, and on and on.

2. ASKLEPIOS VS. HYGEIA

Science is the framework of medicine these days. But physicians are still in touch with the idea of patron saints or gods. All of them, in one form or another take the Hippocratic oath, swearing to do as little harm as possible.

They are less aware of medicine and curing and their ideas are at some odds with the notion of the maintenance of health *within nature*—the complementary views held by the Greek god Asklepios (contra pathology) and his daughter Hygeia (toward maintaining health). Both views existed within the West, but Asklepios' vision of medicine as curing has surely triumphed. In the East, Hygeia is the shining star.

For the worshippers of Hygeia, health is the natural order of things, a positive attribute to which men are entitled if they govern their lives wisely. According to them, the most important function of medicine is to discover and teach the natural laws which will ensure a man a healthy mind in a healthy body. More skeptical or wiser in the ways of the world, the followers of Asklepios believe that the chief role of physicians is to treat disease, to restore health by correcting any imperfections caused by accidents of birth or life.[1]

In the West, especially since antibiotics and many quite amazing technologies, the views of Asklepios have risen. "Scientific" medicine has triumphed over other approaches. Today, however, things grow more complicated. Many of the recent immigrants to the U.S., live more intimately with Hygeia, and have brought their patterns of living with them.

Besides, Hygeia is more frugal, that is to say, *cheaper* than many of the important technological and surgical techniques which have developed in the past several decades. For older people in particular, the question of maintenance of reasonable health and lifestyles with increasing age, does not seem well addressed by scientific medicine.

But the issues surrounding the "scientific" and the "spiritual" in medical thinking and practice need considerably more exploration. Our culture derives principally from European ideas and settings. In it, there seem to be three forms of medicine, two of which remain within medicine as such. The third, which was formerly taught and practiced "as medicine," has moved to what I'll call fringes of a scientific approach to curing.

Most of the medicine practiced in America now is "allopathic"— literally, "against disease." All MDs are allopaths as are almost all our hospitals. But there are other physicians licensed to practice and to prescribe all available medicines. Osteopaths are trained in osteopathic medical schools, have an OD, and are usually not welcomed in most medical schools or allopathic hospitals. Their training, from what I have been told, is much the same as allopathic physicians but they also get trained in the "manipulation" of the body. Parenthetically, *touch* is one of the current issues in medical practice. Most MD's rely increasingly on instruments to diagnose, and scarcely touch patient's bodies! They are, perhaps, out-of-touch.

Much of the difference between allopathy, osteopathy, and the third form of medical teaching—homeopathy—derived historically from the geographic place of the practices. Allopathy was the form of most of Western European medicine, homeopathy was practiced in the northern European countries, including Scotland. It treats disease not against, but *in the direction* as they say of the pathology: e.g., sweat baths for fever. Until allopathy triumphed in America, homeopathy was widely taught and practiced. For example, the U. of Minnesota Medical school taught homeopathic medicine until about 1903, reflecting the Germanic and Scandinavian roots of much of its immigrant population.

The last remaining homeopathic medical school was closed not so many years ago and, now, none remain. But the interest and use of homeopathy remains strong, and much of its increasing practice has moved to the "scientific fringes." In the home with children, homeopathic thinking and practices are quite common. As we explore the relationship between scientific and spiritual, it is important to remember that much of how we think about sickness and curing is learned and practiced at home while we are growing up. We continue to learn and practice it in dealing with our own children's symptoms, trying to decide what is serious or not so serious. To be sure, we practice it on ourselves, attempting to judge how healthy we are at any moment, and what, if anything to do about particular symptoms. Hygeia sneaks into our thought and behavior in all sorts of interesting moments.

3. THE LOCUS OF PATHOLOGY:
INDIVIDUAL OR COMMUNITY

The general scene of medicine is even more complicated in the context of plagues—SARS, AIDS and public health issues generally. These reflect Hygeia's influence on our thinking and practice. We are concerned with the questions that arise about populations and contagion and about how such communal matters of health are to be dealt with. This has been a major issue throughout history, and in America for the past century at least. Sinclair Lewis' *Arrowsmith* poignantly portrays the issues in the early part of the 20th century, highlighting the tensions between allopathic medicine and public health needs, particularly in the context of the allopathic medical school. Here at the University, it was only in the past couple of decades that the Dean of the School of Public Health found an office in its medical school. The relationship between the School of Public Health and the Medical School are more likely to be tense than conversational.

Several decades ago, medicine attempted to co-opt public health by establishing departments of *preventive medicine*, thus, it was thought, keeping public health at bay. But public health issues and needs rise dramatically in moments of terrorism, of epidemic, etc. in our time in short. Ironically, as we look back, the introduction of antibiotics tempted us to think that plagues and epidemics were a thing of the past. AIDS was a rude awakening!

The situation grows more and more clouded and complicated. Does pathology apply to each individual as such. In that case, a person's symptoms should direct him or her to a physician. Or, is the community the effective locus for dealing with some disease entity? The issues of public health are very different in kind from those of allopathy. Moreover, the thinking of most epidemiologists is more kin to statisticians than to physicians. In a time of international terrorism, their epidemiologist's role has increased dramatically.

Our usual ways of thinking about medicine and pathology are challenged. The upshot is that in moments of terrorism or plague, the priority claims of individual pathology are suspended. At the same time, fear propels many people into the wishful and spiritual. Epidemiologists are amazingly effective at creating a climate of fear. Often justified, it also succeeds in encouraging people to wander into the worlds of fantasy. I recall Michael Osterholm, at the U. of Minnesota Medical School, one of the principal epidemiologists in national terrorist policy thinking. He is quite powerful in presenting the worst imaginable scenarios. "Did you ever think," he asks, "about the outcome if a terrorist with smallpox walked through the Mall of America (one of the largest shopping malls in the world), infecting thousands?"

4. THE VARIOUS PRACTICES OF ALLOPATHIC MEDICINE

Further complicating the issue of what is scientific about medicine, is its various specialties, methods, and organizations.. Internal medicine and its sub-branches may be thought to be the clear center of allopathic medicine. Surgery has a radically different history with its quite radical approach to medicine, i.e. a smaller hurt for a greater good. Surgery is a form of deliberate wounding. Obstetrics is allopathic medicine only in the event of difficulties. An ordinary delivery is more a celebration than pathology. Neurology offers few cures. Pathology is analytic and laboratory-based and so usually removed or remote from cure as such.

The "basic sciences" are often ancillary to medicine, per se. Any given science is taken more or less seriously depending on the style and interest of different eras. Physiology, for example, at the U. of Minnesota, was threatened with extinction recently. Anatomy has become a necessary skill for medical students, but it is hardly central to the practice of medicine or curing. A field of anatomy which derived its impetus from space exploration of zero gravity, *functional* anatomy, has become very important to plastic surgery, and to orthodontia in dentistry. On the other hand, it has had very little impact on the general practice of medicine.

Functional anatomy focuses directly on the body moving in gravity. By contrast, most of current medical practice and thinking, perhaps learned from dealing with the cadaver in an introductory anatomy course, seems to extend a deep sense of passivity to the living body. Questions of anatomy like maintenance of posture with aging, are increasingly important to the current patient population but don't seem to have shaped medical practice or thinking very much. Questions of strength, exercise, stretching, and maintaining the body reside more in physical education than in medicine. Except for sports medicine which is driven by the profitability of treating sports' injuries, functional anatomy seems to have very little resonance in medical thinking or practice.

This is not to underestimate or under-appreciate the "advances" in medical thinking and practice that have been made. But what is scientific or not varies quite a lot from practice to practice. Surgery is a very high-level physical *skill*, while much of the practice of internal medicine has to do with observation and with experience, principally having to do with diagnosis. What precisely is scientific about this is not always as clear as we might think. It is more apt here to think of the so-called *art* of medicine.

As far as I can see, the attribution of *scientific* has much to do with the question of *efficacy* and far less to do with posing scientific questions. This is not to say that most research isn't done with very great care. Much of what is more clearly scientific seems to focus on the use of *double-blind* tests of drugs. In such tests, a given drug or treatment is more efficacious than no treatment, or a placebo. Neither patient nor experimenter knows what is being given. Only recently has the question of how or why a placebo should *ever* work to effect a cure been raised in a serious way.

Psychiatry, which has in recent decades turned from "talk therapy" to what is primarily a pharmaceutical activity, still remains somewhat of an aberration in medicine. Many physicians remain leery of psychiatry. Partly, this is due to the fact that in mental disorders the boundaries between science and spirituality often blur.

Demand for advice and treatment concerning mental symptoms or difficulties which seem more mental than physical has certainly increased. The field of psychotherapy, however, has moved from medicine where it was located when I worked in it during the 1960s to psychology and social work, and sometimes it seems to whoever will deal with various complaints and symptoms. In this arena, spirituality and even mysticism often flourish: seeking for advice from the deceased, looking for Jungian ideas which are innate, playing with Tarot Cards, seances, to shamanistic ideas, voodoo, and the like.

5. THE PERIPHERIES OF MEDICINE

Medicine utilizes a number of ancillary professions—or perhaps not quite professions—to assist it in its curing practices. Nursing was certainly the first of these. Among other things, it raises gender issues and calls attention to the fact that medicine is status-driven. Women typically assist doctors, often its seems without paying much attention to who has the most or the best knowledge in the particular case. While women have entered medicine in greater numbers in recent years, nursing remains essentially a women's profession and nurses generally remain helpmates to physicians.

Medicine essentially yields the mouth—a central aspect of our being and principal entry to our inner being—to dentistry. I attended seminars in *oral medicine* for several years, and got a pretty good sense of the boundary areas and skirmishes between dentists and physicians. They seem not only to have different anatomical foci, but seem to think somewhat differently about disease and cures. Much of dentistry is directed toward maintenance of health, and is a field filled with techniques and technologies. Internal medicine, by contrast, is principally situated in diagnosis and hasn't all that many techniques beyond drug therapies. Dentistry also has lower social and professional status than medicine.

The feet are another area of our bodies of central importance The body, after all, lives in and depends on gravity, standing and moving. But medicine essentially yields the feet to podiatry, distinctly less prestigious than medicine. Podiatry uses techniques which are somewhat different in focus from usual medical procedures and thinking. Again, arenas are opened up where curing can easily wander from the scientific to the spiritual.

In the Chronic Disease Institute, I was introduced to physical and occupational therapists. Physical therapists are now being used more ever before, to deal with many of the strains and hurts of being bent out of shape: being hurt, aging, etc. Again, body issues are yielded by allopathy to ancillary workers whose professional status is lower. Rehab gets a little attention. Exercising to maintain health is quite peripheral to medicine, and not taken very seriously by it. This opens the door to quackery, a willingness to take any advice that will help relieve symptoms. Occupational therapy is more clearly situated in chronic disease settings, where patients who have had permanent changes in their bodies, get assistance in negotiating their worlds.

Nutrition—a recurrent, and for many a central issue in life— remains located in the School of Agriculture at the U. of Minnesota. It is located several miles away from the Medical School; and is conceptually, light years distant from it. Hospitals may hire a nutritionist, but medicine as such gives it almost no attention. Strangely, while millions of dollars are spent and thousands of vitamin preparations are taken in our marketing approach to health, people have few reliable places to turn to for or trust for advice.

Exercise remains in the field of Education. Low, even in the low prestige College of Education, understanding and caring for the body are not high up as goals of education. We have forgotten the classic dictum: a sound mind in a sound body. Medicine doesn't seem to care, however, unless and until a pathology is presented.

Seemingly ancillary—but by no means unimportant—is the fact of the pharmaceutical industry which has combined into larger and larger monopolies. In many ways, often not explicitly acknowledged, it sends us advertising messages that shape what we tell to and ask from our doctors. At the same time, it strongly influences what our doctors will prescribe. We have become, in some strange ways, physicians to our own maladies.

6. CHANGING ECONOMICS: CUSTOMERS, MARKETING

The current climate of health care is driven increasingly by an aging population. Many older people are on limited incomes, and are eager for the cheapest ways to a cure. In this rapidly growing group, the sense increases that medicine is less open to them than what they already have in their memories. This, I suggest, contributes to the changing view of what constitutes a cure.

Many older people look less for cures than for relief, for something which sustains or maintains them at varying and often declining levels of health. Much of what is relocating the boundaries between science and spirituality is shaped by this larger group. Thus far, most of them still take all those pills provided to them by physicians, but. . . !

Some level of cynicism comes with increasing costs and the drive by drug companies to market more and more products. This is to say that the present economic climate is moving the patient population to question the basis of medical thinking and practice and to lessen their regard for medicine and the physicians who practice it. More generally, this cynicism spills over to include the sciences and technological progress.

All of this is complicated by the shift from private practice—the doctor-patient relationship—to HMOs where the physician is more like a hired hand than in control of the world of curing. Many of the higher order treatment decisions are made by managers who are nonphysicians. The quality of care often seems and often is driven as much or more by cost-benefit judgments than by medical decisions.

The relationship between physician and patient has been altered to *server* and *consumer*. Remnants of the *sacred* where one *yields* one's body to a doctor in return for a most responsible and thoughtful healing relationship, is decreasing. This is similar to developments in many fields like teaching and even the ministry. Students are more like consumers these days and the mood among students is *buyer beware* rather than an inspired and enduring relationship with their teachers. All of this leads to a diminution of the status of the care-giver, and lessens the importance of the status as well as the claims of scientific medicine.

Yet another issue in the current climate: is the search for the "fountain of youth." Medicine, we presume, *will* solve or eliminate disease and soon we will all live forever. To be sure, progress has indeed been impressive, even astounding, in various areas. Nearly all of us live longer. But quality of life issues remain extremely difficult, and are typically addressed in the context of pathology, i.e. the assumptions of allopathic medicine. The actuality of the aging body is hardly attended to, and a principal reason why alternative medicine is so appealing. Again, the buyer determines the nature of medicine!

7. SOME ASPECTS OF ALTERNATIVE MEDICINE

At least some aspects of alternative medicine become bit more sensible or understandable, if not exactly scientific, when placed within the realm of Hygeia. Organized medicine does come close to Hygeia as interest in preventive medicine grows—motivated in no small measure by real or alleged cost savings. The catalogue of healthy practices includes: eat well, good posture, exercise, movement, have a good attitude, deal with fears, stay out of trouble, find and keep good friends, be honest, be nice, control your sexual urges, etc. These virtues are often given with not a little dose of preachment.

Again, my question about the lines between the scientific and the spiritual? Can praying improve one's health? Does invoking the deity or Gaia or some great shaman or a mystical power make any difference to one's health? In this context, I think too of my continuing exploration of *body work*. I am trying to describe what I am doing and what is happening in *rational* terms. This includes more recent explorations of my aging body by taking violin lessons. I find that I am improving rather markedly in performance and I've even arranged to perform Mozart's 3rd Violin Concerto in G this next fall.

I'm reminded of John Dewey. As we know, he studied with F. M. Alexander, developer of the "Alexander Technique" after he had experienced a most painful upper-back/neck pain at age 59 or so. As a result, he also wrote introductions and appendices for Alexander's books. We also know that Dewey stayed pretty healthy until his death in his early 90s and that he was convinced that body work was very much a part of his being alert and healthy to the end.

No one would deny that we ourselves enter as an agency into whatever is attempted and done by us. That is a truism. But the hardest thing to attend to is that which is closest to ourselves, that which is most constant and familiar. And this closest 'something' is, precisely, ourselves, our own habits and ways of doing things as agencies in conditioning what is tried or done by us . . . the one factor which is the primary tool in the use of all these other tools, namely ourselves, in other words, our own psycho-physical disposition, as the basic condition of our employment of all agencies and energies, has not even been studied as the central instrumentality."[2]

Study of much of how we exist has been minimal, in part at least, because much of our being is reduced to habitual behavior and these are primarily out-of-our awareness. My anthropological-linguistic training suggests this notion to me arising as it does in the effort to elicit the phonemic rules for understanding the sound structures in terms of which the world's languages are constructed. Most people have very little awareness of how they form speech or significant sounds in their native languages. For them as for us this muscular activity is background, is habit. My own body work for some twenty years now has proceeded from Alexander, to t'ai chi, to yoga, and a bit of Pilates.

Alexander Technique is a *rational* approach to one's body. A teacher or mentor assists in exploring one's body and this becomes increasingly interesting and complicated as the exploration continues. A human skeleton is always present in the room to help give one some sense of one's being a body. For example, the dimension and size of the vertebral column, and each vertebra is surprising to most people. It's illuminating to see the skeleton while feeling one's body and looking at one's hands and other parts. Usually one is lying down but not always. Thus, when I began to restudy the violin, I went, fiddle in hand, for advice in working with my body, and the violin and bow. Beginning with my feet, and the general support of the body, I was able eventually to move my bow arm with gradually increasing fluidity. I avoided straining my left side which supports the violin and I learned to *relax* while trying to maintain the postures necessary for what violin play demands.

Much of the Alexander Technique is devoted to exploring our bodies in gravity. This is also true with yoga—I've done a form of Hatha Yoga for ten years or so. Both Alexander and yoga explore the parts of our bodies, teach relaxation, i.e., to let-go of the tensions which we bring to our daily lives. I'm not sure what it means but Alexander usually proceeds from the toes and feet to the head, while yoga goes head to foot.

In rational terms, both techniques attempt to bring the person into the *present moment* of his or her body-in-gravity. While we concentrate on our bodies, we try to *let go* of all the thoughts of past and future, and move into the *present* of one's body. This notion of the physical and intellectual present is central to health. From the point of view of both yoga and Alexander, a great deal of *unhealth* has to do with a *loss* of the present.

T'ai chi works a bit differently. It concentrates on the *balance* of the body in gravity. We are, after all, two-legged creatures whose balance is essential to standing and to moving. How we come to know and use our balance is crucial to our being: much of how our bodies develop is about being and moving in gravity. How developing children learn this remains unclear. I suspect that at least some of the developmental problems we find have a good deal to do with how the child conceptualizes his or her body: where its effective center-of-gravity is, etc.

That being-in-gravity is central to our being itself may be noted by observing how many older persons whose bodies may be said to succumb to gravity, lose balance and walk with a tilt very similar to the post-one-year olds who are just learning to deal with gravity. Thus, many of the complaints of aging for which people seek help fit naturally into the kinds of issues which concern practitioners of alternative medicine. Much of what I observe here has to do with maintaining health (Hygeia) principally by keeping one's body in balance. Yoga and t'ai chi also have more vigorous exercisings and stretchings of the body. The various positions of both attempt to get at its complexities. In our life dance with gravity, when the parts of the body don't move well, they tend to tighten up, to lose elasticity, in other words to age.

All this may seem to border on the mystical and spiritual. Breathing well while moving is crucial to all these studies. Many medical issues, they claim, are due to breathing not very well: too controlled, tightening various parts of the body as we breathe. Even closer to the borderline between science and spirituality is learning how to *breathe-into* various parts of the body. This involves imagining that the breath can go into the body as we control or want it to. This seems to mean that by conceptualizing the breath, we can use this control to find and move the tissues in various parts of the body. While I may not be able entirely to explain it, I think I can teach anyone how to do this in a few minutes—if he or she is open to examining his or her own body. The further claim that a therapist or shaman can move or control our bodies is marginal. I'm not sure, but what I think happens is that the movement of another person's hand suggests to us to move our own body parts. Some of us do this easily; others not at all.

Related questions involve larger issues of the anatomy. We know, from functional anatomy, that our bodies are very dynamic. That is why we try to avoid staying prone in bed for very long or why new mothers are up almost immediately these days. Other traditions' theories of the body lead to practices like acupuncture which is far more this worldly than spiritual. How the connective tissue interacts across the body is still an open question. Various pulses and not just heart beat move through the body. My osteopathic colleague works a great deal with these. But, because anatomy is regarded a finished subject and left unexplored by allopathy, these various possibilities are seldom explored. Truth or mysticism, it's difficult to tell.

Claims about the healing power of prayer raise many questions about how our bodies are and maintain themselves. Remaining in a good mood means that we—our bodies—are at ease: No doubt, optimism has something to do with feeling good, treating our bodies well, and lowering blood pressure. It doesn't seem too far a leap to think that if someone believes in a deity caring for, responding and to, and inspiring them, that they might breathe more calmly, or have feelings that keep him or her healthier than otherwise.

All of these efforts loosely collected under the name of alternative medicine trace back to the individual who maintains good health. Missing in the allopathic model is how we do this. By opening up many of the ideas at home with Hygeia, ideas more likely to come from South and East Asia than the West, new possibilities appear for maintaining good health. To be sure, not all these possibilities are useful or sensible, not all useful or sensible for all persons. Hence, exploration and health! Relax, breathe, come into the present of the body in gravity; stretch. Get in shape, stay in shape. Entertain the optimistic notion that a person can maintain good health better than he or she has in the past. All of this suggests to people that whatever works, may work for them too.

It was Nietzsche I think who pointed out that most of the ideas we call philosophy were crafted by those who were ill and/or aging. Even a humanist may look for relief, cure, or the fountain of youth from which he or she may still sip.

NOTES

1. Rene Dubos (anthropologist-philosopher). *Mirage of Health: Utopias, Progress, and Biological Change*, New York: Harpers, 1959. (110-11) Borrowed from Andrew Weil, *Spontaneous Healing*, New York: Fawcett Columbine, 1995. (4)

2. John Dewey, "Introduction" to Frederick Matthias Alexander's *Constructing Conscious Control of the Individual*, New York. E.P. Dutton, 1923. (xxxii)

Excursions into History

6.

Humanitas and the
Human Genome

A Guiding Principle for Decision Making

Faith Lagay

Genetic scientists predict that we will one day be able to ratchet gene expression up and down rheostatically, not only to evade inherited illness and disability but to alter physical appearance and function, retard the aging process, enhance cognition and perhaps even talent, and shape personality and behavior in as-yet-unknown ways. Although distant, the possibility of such technological capability compels us to ask many questions. Will parents have unlimited freedom in selecting traits for their children if they have the financial resources to do so? If society supports gene therapy to eliminate serious inherited diseases but not elective enhancement, how will we avoid creating a "genobility" of the rich?[1] And, most challengingly, what characteristic attributes of current *Homo sapiens*—from our physically vulnerable bodies to our self-determination, qualified as it is by chance—will we decide to protect and preserve, and on what grounds will we defend and justify the values that underlie those choices?

Geneticists, philosophers, and social scientists have been discussing for more than a decade whether, through modifications like those mentioned above, genetic engineering can change what it means to be human. The answer, of course, is that it can. Transgenic (i.e., mixed-species) animals already pace about in laboratory cages. The

109

same technology can create a physiologically nonhuman being, so defined by the taxonomic standard of its inability to reproduce with other members of its former species. As for altering what it means to be human, if we render our flesh invulnerable to damage, disease, aging, and chance we will have altered the human experience. Futurists foresee that humans will defeat the aging process by becoming more and more bionic until our brains alone are made of DNA, reproducing (exchanging genetic material with other brains) and evolving as command centers housed within durable, inorganic "bodies." The cyborgs are clamoring their way from science fiction into science fact.

From the earliest mention of gene therapy and, especially, genetic engineering, many of those opposed to a cyborgian future have argued against any and all genetic manipulation. Don't start, they say. Do not manipulate genes at all, because, once we start, there will be no stopping. This argument shares a common conceptual structure with other members of a class known as slippery-slope arguments. "Slippery slopers" argue against action or practice A, not because A is, itself, unacceptable or undesirable but because doing A will lead to a consequent action or event B, and B is unacceptable or undesirable. In the case at hand, slippery-slope arguers plead that we not attempt to eliminate inherited disease or fortify the immune system by modifying the genes in embryos, sperm cells, or ova because, if we do, rich people will soon be having designer children, mad scientists will be creating half-human chimeras, and, before you know it, cyborgs.

Slippery-slope arguments are popular and seductive—and harmful. They harm moral and ethical deliberation in at least two ways. First, in their efforts to prevent us from taking the acceptable step A, they take our eyes off the real issue—where is the point between A and B where the action, practice, or event stops being acceptable and becomes unacceptable? Thus, they keep us from attempts to hammer out agreement on difficult value-laden issues. Their greater harm is their logical dishonesty, their claim that the connection between A and B is a necessary one and that humans have neither the will nor the desire to stop at the brink of the unacceptable. Rejecting slippery-slope arguments, on the other hand, opens the possibility of focusing our attention on the pursuit of ethical knowledge through deliberation.

Given the complexity of individual interests and the plurality of cultures that comprise our global society, it is safe to say that no single value system will emerge to guide all deliberation concerning what we want the terms human nature and the human experience to represent in the future. We will use, and need, many systems in answering those questions. My search for a coherent value system to defend my own species-regarding, aesthetically motivated "No" to a robotic future has led me to the humanist tradition and the values captured by the term humanitas as understood by the ancient Greeks, their Renaissance descendants, and contemporary humanists. Humanism provides, in fact, both a rich tradition for valuing and evaluating human nature, and a means for deliberating and decision making in the absence of certainty.

For the Greeks of the fourth century BCE, humanitas meant that which is distinctively human, and that, in turn, meant (whether rightly or mistakenly, according to twenty-first-century knowledge of other species) a capacity for "fellow feeling" realized through education. This central tenet of humanism celebrates compassion as a good and education as a means toward actualizing that good. Embedded in this concept is an understanding of the nature-nurture interaction that still invites much speculation and investigation today. Human feeling was inextricably tied to education by the ancients, according to Renaissance scholar John Stephens, because it is through knowledge that we come to understand the nature of our species.[2] Tzvetan Todorov discusses humanism's view of the interplay between human nature and human nurture in terms of the "given" and the "chosen."[3] The "given" is the irreducible humanness that we share genetically. The "chosen" is the exercise of our will. From humanism's recognition of this ongoing dynamic comes its characteristic faith in education, Todorov says. He continues, "Since, on the one hand, man is partially undetermined and moreover capable of liberty, and on the other, good and evil exist, one can become engaged in that process which leads from neutrality to good, and is called education" (p. 38).

The word humanism was not employed to refer to the tradition I am tracking until the fourteenth century CE, but humanism's progenitors were the Greek rhetoricians who set themselves apart from the philosophers of the time by championing the role of education in preparing free men for self-government. Education trains us to reason.

Reasoning had both instrumental and intrinsic value for Greek educators, the latter stemming from the concept that, insofar as humans were the only species to possess logos (understood as both "logic," the ability to reason, and as "word," the capability for language), we were at our most human when exercising reason.

Other human qualities and capacities the humanist tradition has extolled throughout history are:

1. Appreciation for the human body—its poise, beauty, athletic prowess;
2. The desire for self-knowledge, and the role of the emotions in achieving it;
3. The exercise of artistic creation and its ability to excite the imagination;
4. Valuing and judging itself, that is, the ability to discriminate; and
5. Confidence that rules for ethical conduct can be derived from the affairs and interests of men (in their words) without need for recourse to revelation or divine intervention.

Central to so many of these activities—to education, the disciplining of the body in athletics or dance, to artistic expression, as well as to ethical conduct—is striving. Some reward-feedback system produces great satisfaction in most humans from achieving as the result of striving. No doubt humans who achieve without striving contribute as significantly to society as those who strive, perhaps more so. Yet pleasure redounds to the individual from striving and achieving. We continue to attempt that which we cannot yet do; we work to perfect that which we do satisfactorily. Striving is part of what it means to be human. Consider the many science fiction tales where humanlike organisms have all their needs met—from nourishment to pleasure— and perform their designated functions without striving or stress. Then come one or two who demonstrate dissatisfaction merely because they want to discover something besides the comfort of what they can so easily obtain and do. They want to strive for something not "given." *Gattaca* and *The Truman Show* are recent examples.

Preserving for each human the opportunity to strive, to pursue his or her vision of the good, will not be easy. The temptation will be great to design humans for the tasks the designers would like to see accomplished. First, perhaps, to alter the human body through genetic engineering so that it can survive the temperature and pressure of space travel within our solar system. Later, some humans will desire to go beyond the solar system, a trip that will require indestructible, inorganic bodies and more than one generation of brain. A decision to create humans for specific tasks is a bright line on the slope from A to B, from acceptable to unacceptable. Many genetic changes that leave human self-determination intact and permit choice in the pursuit of happiness fall in the acceptable range. Any genetic manipulation that limits human desire or ability to define and pursue one's own version of the good—one's striving—will depart from a humanist understanding of what it means to be human. Individual parents-to-be will face the test of balancing between the given and the chosen in a microversion of the larger society's challenge: "Am I using genetic engineering to 'give' my child the opportunity to become whatever he or she 'chooses,' or am I using it to create a concert violinist?"

Besides its touchstone set of values, humanism also gives us a method for making decisions about these values. It is the tradition of argumentation or deliberative rhetoric, which developed alongside and in opposition to Athenian philosophy and its masters, Socrates and Plato. These philosophers sought absolute, universal truths through contemplation. The Greek teachers of rhetoric sought not ultimate truths, but a procedure for reaching the best judgment given the particulars of a specific situation. They did not attempt to prove that a given decision could be derived from an absolute good or traced back to a universal truth. They were skeptical of the ability of the human mind to discover truth. Though some opinions might be true, they doubted that the human mind was able to know the difference between those opinions that were true and those that were not.

It is of crucial importance that the rhetoricians' epistemological skepticism did not lead to nihilism—denial of the possibility of the existence of truth, or denial of any foundation for ethical decision-making and moral action. For the missing unachievable proof of truth, deliberative rhetoric substituted consensus, on the belief that in the uncertain human realm of probability and contingency, uncoerced consensus was as good a result as one could achieve—and as good as was needed. Consensus is, itself, an ideal, one we must strive for through education.

Earlier I mentioned the humanist tradition's valuing of emotion, and I want return to that. In *The Art of Rhetoric* Aristotle says that "emotions are those things by the alteration of which men differ with regard to those judgments which pleasure and pain accompany."[4] That is difficult syntax. What it says is that emotions are what cause men to differ in their judgments about what is painful or pleasurable; what they like and don't like; what, in sum, they consider to be happiness and the good life or the opposites of these. The classic Greek trage-dians and comedic playwrights understood the value of emotion in causing us to recognize our own vulnerability and our membership in the community. Philosopher Martha Nussbaum focuses on the role of emotion in at least three of her works.[5] In *The Fragility of Goodness*, she comments that emotion-generated self-knowledge confers prac-tical wisdom, tolerance, and compassion that influence the way the individual so enriched judges other people and situations. A mortal being, she says, "is neither a pure intellect nor a pure will; nor would he deliberate better if he were."[6] The everyday world of deliberation concerning probable outcomes depends upon judgments born of reason and emotion working together.

I returned to the importance of emotions for two reasons. The first is the relationship between body and emotion. Many emotions do not arise simply from bodily pleasure and pain, but many do. Our physical pain makes us feel bad and our physical pleasure puts us in a good mood. We find food and hot baths emotionally as well as physically satisfying. Emotions are, cognitive science tells us, products of com-plex interactions between the brain and the body that find expression in the "mind"—that emergent, dynamic, integrated snapshot of how the brain-body is faring at a particular moment. Cognitive science has not yet entirely figured out the mind. But we know enough about the body's input into "mind" to realize that the range of emotions would sustain a significant blow with the loss of the brain's embodiment. Pleasure would diminish also with the loss of the human body in its infinite variety as the object of our gaze and the subject of our art. This argument for the retention of infinitely varying flesh-and-blood bodies over assembly-line robotic bodies also may ground a decision that genetic engineering should stop short of creating every flesh-and-blood body to a narrow range of specifications—height, weight, shape, musculature, coloring, and so on. Restricting variation seems to restrict response. I hope we think more about variety and diversity as we conspire to achieve ideals of human presentation.

The second reason that I returned to emotions is to argue that they not be banished from deliberation about the future of Homo sapiens. Why should we attempt to eliminate those "things, by the alteration of which men differ in their judgments" about the good life? That is, after all, what we will be coming together to decide. Moreover, we cannot suppress value and emotion from our argument even when we try. Attempts to do so cause us to parade those irrepressible emotions falsely and to call them "reasons."

This defense of emotion brings me back to fellow feeling, love—another factor in the undoing of many fictive societies of the future. Everything is going well among the species we might call Homo anhedonicus, until two among the many discover special feelings toward each other and that's it. It is love, not star-wars weaponry, that topples the emotionless society. The strength of the emotion empowers the two who have discovered it to overcome all sociopolitical restraints and escape from the prison of nonfeeling. By definition, it takes at least two to discover and arouse "fellow" feeling; humanitas is not self-love. So it was recently in *The Matrix*, and before that in *Brazil*, *The Time Machine*, *Fahrenheit 451*, and so on.

I have not spoken of self-determination and can do so only briefly, important as the topic is. Humanism's reliance on education is, of course, grounded in a belief in self-determination. One may train an animal or program a computer that does not have free will, but one does not educate either. If the expressions of our free will, i.e., our choices, are the sum of so many tributary influences and midcourse changes that they are impossible to predict with certainty and often impossible to reconstruct even after we have made them—if, in other words, choice is a matter of "for-all-practical-purposes-free-will," and I believe it is—then we must rely on educating people to evaluate, integrate, and manage all the tributary, and in themselves insufficient, determinants. And we must teach them how, having performed those operations, to act purposively to achieve the desired end. Any genetic step we take that renders the evaluating, integrating, managing, and purposive acting inoperative crosses the line between acceptable and unacceptable.

Though we may eventually accept or reject the values placed by the Greeks on the human attributes and capacities named above, they undergird what appear at first glance to be mere preferences. They provide bases for arguing to preserve education, embodiment in a living, destructible body, emotion, imagination, self-discipline, and self-determination. The humanist tradition also provides a means, an approach, for beginning to reason and feel our way toward making choices as a society for the regulation of genetic engineering.

NOTES

This article appeared in *Free Inquiry*, Winter 2002/2003, 23:1, pp. 34ff. Reprinted by permission.

1. The term genobility was coined by Maxwell J. Mehlman and Jeffrey Botkin, *Access to the Genome: The Challenge to Equality* (Washington, D.C.: Georgetown University Press, 1998), p. 98.

2. John Stephens, *The Italian Renaissance: The Origins of Intellectual and Artistic Change Before the Renaissance* (New York: Longman Inc., 1990), p. 14.

3. Tzvetan Todorov, *Imperfect Garden: The Legacy of Humanism* (Princeton: Princeton University Press, 2001), pp. 36–37.

4. Aristotle, *The Art of Rhetoric*, edited and translated, with introduction and notes by H.C. Lawson-Tancred (London: Penguin Books, Ltd., 1991), p. 141.

5. Martha Nussbaum, *The Fragility of Goodness: Luck and Ethics in Greek Tragedy and Philosophy* (Cambridge: Cambridge University Press, 1986*); Poetic Justice: The Literary Imagination and Public Life* (Boston, Mass.: Beacon Press, 1995); and *Upheavals of Thought: The Intelligence of Emotions* (Cambridge: Cambridge University Press, 2001).

6. *The Fragility of Goodness*, p. 47.

7.

Joseph Fletcher Revisited

Mason Olds

"If a lie is told unlovingly, it is wrong, evil; if it is told in love it is good, right. . . . The situationist holds that whatever is the most loving thing in the situation is the right and good thing."

Joseph Fletcher in *Situation Ethics*

Joseph Fletcher who died in 1991 at the age of eighty-six was an ordained priest in the Episcopal church who spent most of his academic career as professor of social ethics at the Episcopal Theological School in Cambridge, Massachusetts. When he retired from the school in Cambridge, he went on to the University of Virginia where he taught medical ethics for a number of years. He caused quite a disturbance in 1966 when his book entitled *Situation Ethics: A New Morality* first appeared. At about the time the debate about his theory of ethics had run its course, he published *Humanhood: Essays In Biomedical Ethics*, which once again created controversy especially among biomedical ethicists. By now, Fletcher had moved his orientation from theism to humanism, and he was interpreting his theory of situation ethics from the context of ethical utilitarianism.

Perhaps, enough time has passed since his death that we can now approach his work with less heat and more light, less passion and more rationality. As we revisit Fletcher's work, there are two fundamental questions to be discussed: first, what is the basic moral principle or principles for judging whether a particular action is moral? It is in answering this question that we see how his theory validates a moral action. Second, why should one accept this principle or principles as a moral action guide? This question is a higher order question than the first, and is much more difficult to answer. It deals with the important problem of vindicating his ethical theory. It is with these two questions in mind that we approach Fletcher's ethical theory.

1. THE ORIGIN OF SITUATION ETHICS

Situation ethics emerged out of the religious, ethical, and social turmoil that came to a head in the 1960s. It is a theory that arose out of Christian ethics. In 1962, Bishop John A.T. Robinson, Bishop of Woolwich in the Anglican church published a book entitled *Honest to God*, which was widely read, discussed, and debated.[1] Bishop Robinson did not say anything particularly original; he simply brought together some of the central ideas of such theological giants as Paul Tillich, Rudolf Bultmann, and Dietrich Bonhoeffer. But the fact that the Bishop said to the lay public what scholars in theological schools had been teaching for two decades caused quite a disturbance in the churches in England and the United States.

Bishop Robinson stated clearly that God is a problem for modern men and women, that previously held notions of God are obsolete, and that a radical revision is necessary if the Christian faith is to speak meaningfully to people today. The Bishop stated the problem but remained somewhat middle of the road in his proposed solution. The "death-of-God" theologians, who had read their Nietzsche, also saw the problem, and went much further in proposing a remedy. In various ways, they concluded that God is absent from the contemporary situation and that the cause of his absence is that he has died.[2]

Interestingly enough, Robinson not only perceived the problem of God but also saw the implications of the problem for Christian ethics. In fact, he devoted a chapter in his controversial book to the "new morality." His remarks were virtually ignored until the Profumo Affair (involving a high ranking government official sleeping with a cold war agent) became public shortly after the appearance of *Honest to God.* As it turned out, the Bishop was first attacked for his religious skepticism and then attacked as an immoral libertine and an advocate of that horrible "new morality."

One of the contributions of *Honest to God* was that it clearly defined a relationship between Christian theology and Christian morality. If God has become a problem for modern people, then morality based on an understanding of God has likewise become a problem. Further, if a prescientific theory prevents people living in an age of science and technology from taking Christianity seriously, then a morality from an earlier age has become a stumbling block. Hence both the "radical theology" and the "new morality" were present in germinal form in the Bishop's book.

In many respects radical theology and the new morality were two sides of the same religious coin. The death-of-God theologians became the most vocal segment of the radical theological camp, and eventually through the further works of Bishop Robinson in England and Joseph Fletcher and Bishop James A. Pike in the United States, a specific type of the new morality came to be designated "situation" or "contextual" morality. For instance, as I have already noted, Fletcher's popular and significant work *Situation Ethics*, carries the subtitle *A New Morality.*

To prevent confusion about the relationship of situation ethics to the new morality, it should be noted that during that turbulent decade of the 1960s the term *new morality* was employed in three different ways. In addition to designating situation ethics, it was sometimes used to refer to such phenomena as "the Playboy Philosophy" advocated by Hugh Hefner, then editor of *Playboy* magazine and president of the Playboy empire. The term was also used to allude to the search within the student protest movements for a new moral perspective. The common element in all three perspectives is the conviction that the old morality was no longer viable for the present situation and that the times demand a new morality. Here, of course, harmony ends and different reconstructions begin.

In Fletcher's attempt to trace the origin of the concept of situation ethics, he pushes back into the early 1950s. He says, "This contemporary shape of Christian ethics was accurately described and labeled as 'existential' or 'situational' by Pope Pius XII in an allocution on April 18, 1953. He denounced it, of course, pointing out that such a nonprescriptive ethic might be used to justify a Catholic leaving the Roman Church if it seemed to bring him closer to God, or to defend the practice of birth control just because personality could be enhanced thereby! Four years later, February 2, 1956, the Sacred Congregation of the Holy Office called it 'the new morality' and banned it from all academies and seminaries, thus trying to counteract its influence among Catholic moralists."[3]

Although I have emphasized the social and intellectual turmoil of the early 1960s, and although Fletcher goes back a decade earlier for the context in which situation morality arose, in a broad sense it can be attributed to Jesus, nearly two thousand years ago. Jesus allowed the context to influence his understanding of right and wrong. If this is true, then situation morality is not new at all but the old morality of the founder of Christianity.

2. SITUATION ETHICS VERSUS LEGALISM

One way of gaining some understanding of a philosophy is knowing what it is reacting against. In this case, Fletcher and the other situationists are reacting against legalism. For instance, Fletcher contends: "Legalism in the Christian tradition has taken two forms. For Catholics, it has been a matter of legalistic *reasoning*, based on nature or natural law . . . Protestant moralists have followed the same adductive and deductive tactics. They have taken Scripture and done with it what Catholics did with nature . . . One is rationalistic, the other Biblicist; one natural, the other Scriptural. But both are legalistic."[4] Often Fletcher refers to legalism as "the old morality," and construes it to be a type of morality, usually theological, that claims to have an absolute and universal standard of right and wrong. So the two major forms of this way of thinking, although there are important differences between them and even among the advocates of each position, are traditional Roman Catholic morality and Protestant morality. Both emphasize immutable laws or principles as standards for moral behavior. In one way or another, moral law was considered infallible; it could not be violated without the results being an immoral act, regardless of the situation.

By way of example, recall Thomas Aquinas. Although he lived in medieval times, he continues to have a profound influence on Roman Catholic morality and illustrates one kind of legalistic morality.[5] According to Aquinas, God created the universe in accordance with his eternal law, which is known only in its fullness by Him. However, part of God's law has been revealed to humanity and this revelation has been recorded in Scripture. That part which appears in the Old Testament is referred to as the law of "the Levitical Priesthood;" that, in the New Testament, as "the Priesthood of Christ." Of course, the Church, through its head, has sole responsibility for correctly interpreting the law. Apart from the Divine law, humanity is capable of discerning part of the eternal law by the use of reason and inclination. This part of the eternal law, which humanity is capable of knowing, is called the natural law. Thus, Aquinas' moral theory, as we have seen, is typically referred to as a "natural law theory."

For example, the traditional Thomist will argue that the primary purpose of sexual intercourse is procreation. Therefore the natural desire of man and woman for each other has been created in them by God, whether they know it or not, in order to produce offspring for the continuation of the species. Any "artificial" method of birth control, whether chemical or mechanical, frustrates the primary purpose of coition. It is thus a violation of nature, which in turn is a violation of natural law. Quite simply, birth control is immoral regardless of circumstance or situation. Although many American Roman Catholics ignore it, this was the rationale behind Pope Paul's encyclical, *Humanae Vitae*, which reaffirmed the Church's ban against "artificial" means of birth control.

The late Bishop James A. Pike, who was an advocate of situation ethics and, interestingly, was once a practicing Catholic, does not find the natural law theory of morality very convincing. Pike argues that the first problem involved with natural law theory is in stating it. He writes: "When natural law conclusions are stated specifically enough to be relevant to a given situation, we are in the realm of the conditioned and contingent; when they are stated broadly enough to be perennial rather than ephemeral, they become platitudinous and uselessly abstract."[6] In other words, no one would argue with a general maxim like "to each according to his due." But the important and difficult question is "What is his due?"

Pike also maintains that natural law morality has been used to support contradictory points of view. For instance, it has been used to restrict religious freedom by maintaining that "error has no rights" (a point of view held by Aquinas himself and many Catholic moralists prior to Vatican II) and also to promote religious freedom by contending that all human beings should have the right to worship and witness according to their own consciences (post Vatican II).

In continuing his argument Pike says that sometimes natural law morality claims to arrive at its principles inductively, that is, from observations that attempt to discover a universal consensus. But the great variety of moral systems emerging from a study of history, sociology, psychology, and other disciplines lead Pike to think it impossible to arrive at such a consensus. On the other hand, if the natural law moralist attempts to arrive at his or her principles deductively, Pike notes (and as any first year logic student learns) "that the conclusion of the syllogism is already built into the major or minor premise. In turn the premise is based directly on faith, or derived from a preceding syllogism, or derived from syllogisms whose major or minor premises are based on faith."[7] After making these criticisms of the natural law theory, Pike concludes, "'Natural law' is simply a holy noise or a color-words phrase used forensically in support of a current position held on other bases. It is fruitless to look here for an absolute grounding of ethical positions."[8]

Situation ethics is also a repudiation of Protestant legalism, another version of the old morality. One of the central tenets of Protestantism is that the Bible is the norm for both faith and practice. Although in fact the Bible has been interpreted in a variety of ways, it is claimed that it has remained constant and clear. In the seventeenth and eighteenth centuries a kind of Protestant scholasticism developed, which tended to straitjacket the interpretation of Scripture. In some sense the Scripture was considered infallible, although there was not always consensus about exactly what that meant. This rigid approach to the Scriptures has continued to the present day, so that many Americans are taught and still believe they have an infallible guide for faith and practice.

Protestant legalistic morality functions somewhat like Catholic morality. The Protestant takes the Scriptures and employs them much as the Catholic uses nature. In other words, the Protestant accepts the Bible as containing revealed truths from God about right and wrong. The usual source for moral judgments is either the Decalogue, which God is said to have handed down to Moses on Sinai engraved upon stone tablets, or the moral teachings of Jesus, usually taken from the Sermon on the Mount. The Protestant legalist converts these revealed truths into divine laws, arguing from them to concrete situations. So the legalist knows what is right and wrong before the concrete situation arises. The answers are preset, and all that is needed is to pronounce the judgment as soon as the situation arises. Both scriptural and natural law legalists thus insist that they have an objective moral standard and, in the past, they have been successful in spreading this view. The public has accepted their judgments.

But Fletcher claims that moral boundaries are more blurred; because moral problems are more complex than the legalists acknowledged. Only the fundamentalist, the evangelical and the orthodox faithful now accept without doubt the answers as unquestionably valid. More and more people challenge the authority of the old morality. Of course, one of the reasons the old morality is not taken as seriously today as before is that it offers absolute answers to problems that people believe are altered by the situation. Furthermore, the churches have changed their position on questions that they had either directly said or implied were absolute. So, people are aware that the churches, rather than providing an objective, infallible, absolute basis for moral pronouncements, are historical institutions plagued by the same provincialism and relativism as other historical institutions. Until the early part of the twentieth century, for instance, both Protestant and Catholic legalists thought that artificial means of birth control were immoral. Hence Protestants placed Comstock laws on the statute books of our various states, and often Catholics kept them there. Today hardly any Protestant will argue that birth control is immoral, and even a majority of Catholics in the United States agree with them. Since the churches, the guardians of the old morality, have changed their positions on this and so many other issues in recent history, it is difficult for the public to take seriously their claims to absolute certainty in dealing with specific moral and social problems. Or as Fletcher says: "Every religious legalism, whether of the Catholic natural law variety or the Protestant Scriptural law variety is sooner or later repudiated."[9]

3. SITUATION ETHICS VERSUS ANTINOMIANISM

Having established one boundary by showing that Fletcher and the situationists deny the validity of legalism, I am now ready to establish its other boundary, i.e. situation ethics as a repudiation of antinomian morality. As Fletcher says: "Over against legalism, as a sort of polar opposite, we can put antinomianism. This is the approach with which one enters into the decision-making situation armed with no principles or maxims whatsoever, to say nothing of *rules*. In every 'existential moment' or 'unique' situation, it declares, one must rely upon the situation of itself, *there and then*, to provide the ethical solution."[10] The term *antinomian* means against the norm or law. Like Fletcher, the antinomian maintains that the individual enters the moral situation without any reliable fixed laws or principles and learns that each situation is so different that it must be judged on the basis of its uniqueness. In other words, rather than entering the moral situation with the answer already in hand, as the legalist does, the antinomian brings the moral solution out of the situation.

Broadly speaking, there are two types of antinomianism. One is religious as exemplified in the ideas of Soren Kierkegaard. He thought it was possible for a person to orient himself or herself in the world in one of three ways. These orientations or "existence spheres" are not only theoretical but also represent perspectives based in his personal experience. Kierkegaard truly believed he had experienced each of these "existence spheres" in his own life. First, one can be an esthete, living life on the surface without deep inward commitments. The esthete responds to situations by doing the things that bring pleasure and avoiding the things that bring pain. Often, however, the esthete becomes bored with life. The superficiality of esthetic existence does not involve very much of his or her total being.

Second, when boredom and meaninglessness set in, the individual is driven to despair, which in turn opens the possibility of a new "existence sphere." Kierkegaard refers to it as the "ethical sphere." To be accepted, it must be consciously chosen. The individual must knowingly relinquish the esthetic mode of existence and makes a "leap" into the ethical sphere. Now the person reorients himself or herself around the ethical demand to do one's duty, thus choosing to live by a universal moral principle or the common good rather than by the limited self-interest of the esthete.

According to Kierkegaard, however, the "ethical sphere" is not the highest form of existence. The ethical individual loses himself or herself in making the commitment to the universal demand to do his duty. When the individual feels that he or she has lost himself or herself in the universal, a third and final sphere, the "religious," presents itself as a possibility. Once again the individual must make a choice; deciding to "leap" from the ethical sphere to the religious. This is the leap of faith, and for it the most extreme risk is present. The individual turns from trust in himself or herself to trust in God. Hence the individual now gains a fuller personhood. The norm for his or her decisions becomes "doing the will of God." Kierkegaard concludes that the individual in relation with God, is superior to the universal call of duty of the "ethical sphere":

> Faith is precisely this paradox, that the individual as the particular is higher than the universal, is justified over against it, is not subordinate but superior—yet in such a way, be it observed, that it is the particular individual who, after he has been subordinated as the particular individual, now through the universal becomes the individual who as the particular is superior to the universal, for the fact that the individual as the particular stands in an absolute relation to the absolute.[11]

Kierkegaard's contention in this complicated passage is that the highest form of existence of which humanity is capable results from a person's relationship to God. When the individual reaches it, then the morally right thing is doing the will of God. Although the individual may enter the moral situation with guidance from natural law or Scripture and with the morality of the community, ultimately these are of lesser moral value. God reveals the moral answer to the "knight of faith." It takes precedence over all other norms. So Abraham, the biblical example *par excellence* in Kierkegaard's view, suspends the universal moral norm against killing and, following the will of God, prepares to sacrifice Isaac, his son, and thus commit holy murder.

As brilliant as Kierkegaard's perspective may be, it is problematic and dangerous. The most obvious question to ask is: how does one distinguish between the word of God and the word of Satan; how does one know the inner voice of faith as against the inner voice of psychosis. After all, many awful crimes have been committed because "God told me to do it." One also asks: is it really possible that the demands of universal morality will be in conflict with the will of God and visa versa? Martin Buber, as against Kierkegaard, believes that the demands of universal morality are in fact the will of God. Implicit in Buber's view is a way of checking out claims to know the will of God. As he notes, "It is doubtful that what God requires universally will be suspended privately." So Kierkegaard leaves the solitary individual dangling before what he or she believes to be the will of God.[12]

However, antinomianism is not limited to a religious type. It also has a secular version as represented by Jean-Paul Sartre's *Existentialism is a Humanism*.[13] Unlike Kierkegaard, Sartre was an atheist. For him, there is no divine will located in sacred Scripture that we can follow. His atheism also maintains the absence of any common human nature or essence. This means, of course, that there is no God to reveal His will to Kierkegaard's solitary individual. It also means that without a Creator, each person is radically unique and is uniquely responsible for creating himself or herself. Each individual creates his or her own essence or the kind of person he or she wishes to become.

Without God, the universe is indifferent to values. Even if humanity should choose to follow values that could be shown to be valid and good, there is no guarantee that good will conquer evil. According to Sartre, individuals must live and act without moral hope, for there is no God behind the universe giving support to that which is good. When an individual finds himself or herself in a moral situation, there are usually several alternative ways of acting. Since there is neither God, nor common human essence, there is no objective moral ground for choosing one action rather than another. Without an objective norm or principle for acting, the individual is left to choose and thereby to create his or her own values. As the individual chooses, he or she assumes responsibility. That is what Sartre means when he says, "man is nothing else but what he purposes, he exists only in so far as he realizes himself, he is therefore nothing else but the sum of his actions, nothing else by what his life is."[14]

Both Sartre's and Kierkegaard's positions are antinomian. Neither provides an objective criterion for deciding what is right and wrong. Both contend that the solitary individual must enter the moral situation without any prior moral norm. For Sartre, the subjective individual must decide. For Kierkegaard, the solitary individual must receive the answer in a direct revelation from God.

The position of Fletcher's situation ethics now comes more clearly into focus. He is a moderate and views both legalism and antinomianism as extremes. Fletcher agrees with the antinomian that the situation will be one of the factors to be considered in determining what is right and wrong, but he disagrees with the antinomian about the lack of a norm. He also agrees with the legalist that there is a norm but disagrees about its exact nature. For Fletcher, the normative principle must be applied to each new specific situation. The answer cannot be known ahead of time; it must come from the individual's confrontation with the moral problem. Fletcher, then, sees both strengths and weaknesses in legalism and its polar opposite, antinomianism.[15]

4. THE BASIC MORAL PRINCIPLE
OF SITUATION ETHICS

Having placed situation ethics in an historical context and differentiated it from both legalism and antinomianism, I am now read to spell out more carefully what it is that situation is *for*, that is, what does Fletcher advocate? As we know, he places considerable emphasis on the situation or context. But a new question now arises: what is the basic moral principle? What norm must be applied to the situation if it is not the law deduced from nature, from Scripture, from a God's revelation, or from a private subjective response. The answer quite simply is love. Love is the basic moral principle for situation ethics. When a moral question arises, the appropriate answer is do the loving thing in that situation.

Situation ethicists of various persuasions will agree that love is the basic moral principle for determining what is morally right in a particular situation. Bishop Robinson expressed it this way:

> Love alone . . . has a built-in moral compass, enabling it to 'home' intuitively upon the deepest need of the other, can allow itself to be directed completely by the situation. It alone can afford to be utterly open to the situation, or rather to the person in the situation, uniquely and for his own sake, without losing its direction or unconditionality. It is able to embrace an ethic of radical responsiveness, meeting every situation on its own merits, with no prescriptive laws.[16]

Joseph Fletcher concurs with Robinson. He write, "Christian situation ethics has only one norm or principle or law (call it what you will) that is binding and unexceptional, always good and right regardless of the circumstances. That is 'love'—the *agape* of the summary commandment to love God and neighbor."[17]

In situation ethics the individual enters the moral situation not with rigid, fixed laws about what is right and wrong but rather with the moral maxims or mores of the community in which the individual lives and with the values of the tradition in which he or she has been nurtured. However, neither the mores of the community nor the values of the particular tradition are to be taken as absolutes. They are to be used as illuminators and can be compromised or set aside if the situation demands it. For instance, Bishop Robinson says, "Law has its place, but that place is at the boundaries, and not at the center."[18] Love is at the center and takes precedence over law.

Although Joseph Fletcher, John A.T. Robinson, and James Pike were Christians, one need not be a Christian in order to be a situation ethicist. Of course, Christians arrive at their understanding of love from their faith commitment, but Christians do not have a monopoly on love. Some might even have a greater capacity to love without a Christian commitment. A person may come to a perceptive understanding and attain a greater capacity to love with some kind of non-theistic, humanistic commitment. For instance, Fletcher says, "Lovingness is often the motive at work full force behind the decisions of non-Christian and non-theological, even atheist, decision makers."[19] Hence it is possible for both the religious and the non-religious human being to advocate situation ethics. Each will argue that situations alter cases and that love is the basic prescriptive normative principle. But each will come to his or her understanding of love from a different perspective.

Illustratively, Erich Fromm was a non-Christian. He thought love should be the prescriptive norm in human relationships. He presented his views in a short but readable book entitled *The Art of Loving.*[20] In it, Fromm notes that there are several kinds of love, among them self-love (not to be confused with selfishness), erotic love, motherly love, fatherly love, brotherly love, and love of God. During the maturation process, the individual is enriched by experiencing each of these kinds of love. In its proper place, each is good for the individual. As the individual matures, he or she will be able to love maturely, that is, he or she will have the ability to assign each of these loves to its proper place and to express the appropriate kind of love in relationships with others. Of course, one kind of love does not necessarily exclude another type.

According to Fromm, love begins when a person feels another person's needs to be as important as his or her own. When this happens, love becomes an active force in people that enables them to break through the walls that separate them. As these walls are penetrated, one person is united with another. Each shares his or her aliveness with the other, for each gives to the other. This unity of individuals is mutually enriching. While unity exists, each individual also retains his or her separate identity.

Four ingredients are important for expressing mature love. (1) We feel concern for the one we love. As Fromm says, "Love is the active concern for the life and growth of that which we love."[21] (2) We feel a sense of responsibility for the person or persons we love. This is a voluntary act, not something that we are forced to do. In accepting responsibility for meeting the needs of the other, we respond. (3) Mature love includes respect for the other. We see the other as he or she really is and are aware of his or her unique individuality. Respect entails that we will not exploit the other, for we want him or her to grow and unfold in his or her own ways and for his or her own sake. (4) Finally, we feel the necessity for knowledge. We must know the other. Without knowledge, concern and responsibility are blind. Knowing the other enables one to pass beyond the superficial relationship. This becomes possible when we transcend ourselves and see the other in his or her own terms. Yet the more we know the other the more he or she will elude us, a fact that paradoxically helps us understand the other's depth.

My point in elucidating Fromm's understanding of mature love is to undercut the popular, superficial, romantic notion of love. Obviously, loving a person maturely is not easy as we notice when we speak of concern, responsibility, respect, and knowledge. Yet it is mature love which the situation ethicist seeks to apply to the situation in order to determine what is right and wrong. The Christian calls it *agape*, the psychoanalyst calls it mature love. In either instance, it is a love that seeks to do what is best for the other, rather than using the other to meet one's egotistical needs. As Dietrich Bonhoeffer said, it is being "the man for others." Mature love enhances the life of the other, helping him or her to grow and to be more fully alive.

5. HOW SITUATION ETHICS WORKS

In order to see how the theory of situation ethics works, let us take a specific moral issue and examine it from the perspectives of legalism, antinomianism, and situation ethics. This analysis will, I trust, make situation ethics come more clearly into focus. Since Fletcher was quite interested in the moral issue of abortion, I shall use it to show how his theory of morality works. Let's consider two cases which he deals with specifically. First, there is a case which involved a Nazi concentration camp during the Second World War. In this camp there were thousands of Jewish women. If a medical report revealed that a woman was pregnant, she was sent immediately to an oven and incinerated. In the camp there was a Romanian woman physician. When she discovered that a woman was pregnant rather than report her, she performed an abortion. In total she aborted the fetuses of three thousand Jewish women, and in so doing, she saved their lives.[22]

A second example involves an event which took place in 1962 in a state mental hospital. In this case a mentally ill male patient raped an unmarried young women patient who was suffering from severe schizophrenic psychosis. As a consequence, she became pregnant. When her father learned about the incident, he charged the hospital staff with culpable negligence and requested that an abortion be performed immediately. The hospital staff refused on the grounds that the mother's life was not in danger.[23]

Both cases are unusual and interesting. They provide us with striking situations in which to raise questions about the morality of abortion. Was it moral or immoral for the doctor in the concentration camp to perform three thousand abortions? Was it moral or immoral for the staff in the mental hospital to refuse to abort the fetus of the rape victim? How would the legalist, the antinomian, and the situationist answer these questions?

Advocates of all three positions will agree on a number of issues, but strongly disagree on others. For instance they will agree that a human fetus is of the species *homo sapiens*, and they also will agree that a developing fetus is alive. However, they do have important disagreements, and because of these differences, they reach different conclusions about the morality of abortion. In order to gain perspective on the various positions, I raise two questions: *what* makes a fetus a person; *when* does a fetus become a person?

In defining the essential element of personhood, people in the abortion debate have suggested three things. One group has suggested that it is simply life. As long as there is life, a human organism is a person. From this perspective, a person comes into existence at fertilization and continues to exist through the whole complex developmental continuum. This view is not as simple as it might appear, e.g. before fertilization, oocytes and sperm are alive and at the other end of life, tissue is not dead when the brain dies. Nevertheless, on this view person and life are co-existent.

A second perspective is that the soul and not life is the essence of a person. Some advocate an "immediate animation" theory which maintains that the soul comes into existence at the moment of fertilization. This appears to be the "official" view of the Roman Catholic Church and of some Protestant fundamentalist groups. Others promote a "delayed animation" theory which contends that God creates a soul and infuses it into the fetus at the time of "quickening." Both Aristotle and Aquinas thought that "quickening" occurred forty days after fertilization in males and eighty days after fertilization in females. Although the Church has been greatly influenced by the thought of Aquinas in many ways, it deviates from his position and takes a more restrictive point of view.

The third perspective places emphasis on cerebration or reasoning as the essence of personhood. To qualify as a person an individual must possess a certain mental capacity. To avoid misunderstanding, feeling is counted as one of its components. The developing fetus does not qualify as a person because it does not yet possess curiosity, affection, self-awareness, memory, purpose, and conscience. On this view, a human being devoid of minimal intelligence is not a person no matter how many of his/her organs are active. In other words, a baby born with only a brain stem, but no brain, would not be considered a person.

As we cans see, people have quite different views about what constitutes a person. Is it life? Soul? Or cerebration? If we could achieve consensus on this question, we would be a long way toward solving the moral problem of abortion. Since such a consensus does not seem possible, Fletcher believed that any attempt to impose a single doctrine on those who do not share it was itself morally intolerable.

Closely related to the issue of personhood is the problem of determining when the fetus becomes a person. Basically, there have been six proposals which need only listing here. They are at (1) fertilization, (2) implantation, (3) heartbeat), (4) quickening, (5) viability outside the womb, and (6) birth after being expelled from the womb, the umbilical cord cut, and the lungs working independently. As with the problem of personhood, there is no consensus with respect to when the fetus becomes a person. Both issues are at the heart of the debate about the morality of abortion.

With respect to the two cases, how would advocates of the three approaches to morality respond? One type of legalist would maintain that what constitutes a person is life itself. It begins at the moment of conception with both soul and body intact. Thus, in the concentration camp both the pregnant women and the fetus have an equal right to life. It would be morally wrong, then, to give priority to post-natal life.

The doctor should not perform the abortions although six thousand people will die, i.e. on this view, both the women and the fetuses. The three thousand women might find comfort in the belief that they will be rewarded in the life to come. They might also be assured that the Nazi murderers will receive their punishment in eternity. But I suspect this would be cold comfort in the situation. Of course, to abort the fetus of the girl in the hospital would be interpreted as an immoral act. Fletcher quotes a Catholic lawyer as saying, "One person's freedom to obtain an abortion is another person's right to live."

Another type of legalist will have more room to maneuver by taking the position that what makes a fetus a person is a rational soul that is created by God and infused into the fetus at the time of quickening. Aquinas, as already noted, accepted this view, believing that the parents created the body through coition and that God created the rational soul. Hypothetically, then, one would not be committing murder if in both cases abortion occurred early in the pregnancy before the time of the infusion of the soul into the fetus. Should this condition be met, a modern Thomist could praise the actions of the doctor in the concentration camp as moral and justify an abortion for the girl raped in the hospital.

The religious antinomian would pray to God to provide guidance as to what to do in these two cases. Since I do not wish to second guess God, I shall remain silent. However, a non-religious antinomian like Sartre would maintain that there was no objective norm for choosing one course of action rather than another. One simply must decide and take responsibility for the decision. If the Jewish women sought an abortion in order to save their lives and if the doctor performed the abortions, Sartre would have no difficulty with their decisions as long as they were done voluntarily and in good faith. Since the girl who was raped was mentally incompetent, the decision for abortion rests with her parents who, in turn, would be responsible for it. Sartre would probably support the parent's decision for an abortion as an appropriate existential choice. Of course, he could equally support the opposite decision if, again, made in good faith.

Fletcher thought it was cerebration or a minimal level of rational ability that differentiated a fetus from a person. He also thought a fetus became a person at birth. When the pregnancy is wanted, he saw it as a healthy process. However, when it was not wanted, he viewed it as a disease. He believed that it was possible to justify a woman's freedom to choose and that it was impossible to justify compulsory pregnancy. The basic moral principle he would apply in the two situations is: do the loving thing. If this principle is followed it will produce the greater good for the people affected by the action.

The Romanian doctor who performed the three thousand abortions did it out of love or compassion for the Jewish women. Since the fetuses are incapable of cerebration and have not been born, they do not qualify as persons in Fletcher's schema. The Jewish women possess both of these characteristics. It is therefore a mistake to speak of six thousand potential murder victims and contend that the doctor murdered three thousand babies in order to save three thousand women. The fetuses lacked the status of personhood which the women possessed. The doctor did a loving thing and, in so doing, saved the lives of three thousand actual persons.

In applying the principle of love to the girl who was raped, Fletcher would approve an abortion. It is not murder since is still in the early stages of development even by the standards of such legalists as Aristotle and Aquinas. Furthermore, the most conservative legalist will grant a person the right of self defense. In this case, Fletcher holds that the girl has two aggressors, the rapist who is legally innocent by reason of insanity and the fetus who is, although unintentionally, also an aggressor. In effect, Fletcher asks, is it not the loving thing to do to terminate the pregnancy and thus to protect the girl against continuing aggression?

6. THE VINDICATION OF SITUATION ETHICS

Having examined two cases in involving the morality of abortion, we have illustrated Fletcher's view of how doing the loving thing in the situation works. However, we must now ask why one should adopt his basic moral principle. This brings us to the problem of vindication or justification.

Fletcher was certainly aware that traditionally in the West there was a connection between religion and morality. Since Nietzsche, for those who have taken him seriously, the connection between the two has not only been sundered as the Enlightenment held, but both religion and morality have become suspect. Thus, for a moral theory to be viable today, it must be able to respond to this historic context. As we live in a pluralistic world, many moral systems take for granted the values of their own community which has cogency only for those who live within it. Those who live in different communities therefore meet each other as moral strangers. This is particularly true of those who live within communities that advocate a restrictive legalism. By contrast, although there may not be many takers of Fletcher's situation ethics, loving is a principle which people who are committed to a variety of perspectives can discuss and debate. In other words, moral strangers can meet on a common ground which is: should people live by the moral principle of doing the loving thing?

Fletcher's basic moral principles invite us to consider four presuppositions which he offers as a vindication for his moral theory. First, he says that situation ethics is personalistic. It is concerned more with persons than with abstract principles or laws. It assumes that laws are created by and for people, not people for the laws. It is especially concerned about a person's relationship with other people, and it is based on the belief that people have certain responsibilities to each other. How they meet or carry out those responsibilities is important.

Second, Fletcher's ethics is positivistic, that is, its presuppositions cannot be proved rationally or in any other way. Instead, they are posited. By making this acknowledgement, he is avoiding what is called the naturalistic fallacy, the attempt to deduce an "ought" from an "is." In other words, he is denying that good and bad are in the nature of a thing, that good is objectively present in the thing or the action. Moral discourse is different from empirical discourse. So nothing is good or bad in itself. A value judgment is derived from the moral principle adopted by the individual for the purpose of making moral judgments. Hence the moral principle is affirmed by choice and not by reason.. But this does not mean that the moral principle is irrational; rather, it is a-rational. In other words, it is outside reason, not against it. Since the acceptance of a principle involves a decision, the individual must make a leap of affirmation before he or she can make a moral judgment. By accepting the love principle, the situationist can apply love to the concrete situation and by so doing, make moral decisions.

Third, closely associated with the positivistic aspect of Fletcher's theory is its relativism. This means two things. First, there is relativism in choosing one moral theory rather than another, and second, there is relativism emphasizing the uniqueness of each situation or context. The individual must do the work of connecting the moral norm to the particular situation. Love in the abstract is empty; it can only become significant when it is related to the particular problem in the specific context.

Fourth, Fletcher's ethics is pragmatic. In order for something to be usefully moral it must be viable or as ethics puts it, "ought implies can." If it will not work, then it is no answer at all. For instance, for legalists to argue that the Jewish women in the concentration camp have no more right to life than their fetuses, for them to know that if they refuse an abortion they will go to their certain death, and for them to be offered the questionable comfort of eternal reward for their refusal simply will not work.. Fletcher might well add, these reasons ought not to work because in that desperate situation, they are immoral.

To sum up, Fletcher's theory of ethics attempts to be personalistic, positivistic, relativistic, and pragmatic. The individual has to make a decision. It cannot be found in some sacred canon. If you agree with Fletcher's concerns and if you believe that he deals with them more sensibly than others do, then you will accept "situation ethics." If so, Fletcher has vindicated his theory.[24]

NOTES

1. John A.T. Robinson, *Honest to God*, (Philadelphia: Westminster Press, 1963).

2. See, e.g., Richard L. Rubenstein, *After Auschwitz*, (Indianapolis: Boobs-Merrill Co., Inc., 1966), p. 224 and passim.

3. Joseph Fletcher, "Six Propositions: The New Look in Christian Ethics," *Harvard Divinity School Bulletin* 24 (1959), p.16.

4. *Situation Ethics*, (Philadelphia, Westminster Press, 1966), p. 21. Fletcher's understanding of ethical theory evolved over his career. In his earlier works, he promoted a theory of personalism, in the 1960s situation ethics and, as a medical ethicist, act utilitarianism. Basically, I am presenting his theory in the language employed in the 1960s.

5. Thomas Aquinas, *Summa Theologica, I-II*. 91, p 2.

6. James A. Pike, *A Time for Christian Candor*, (New York: Harper & Row, 1964), p. 42.

7. Ibid., p. 43.

8. Ibid., p. 45. Of course, there are many Catholic moralists who are trying to rethink morality in the present context. They employ casuistry in quite loving ways. See, e.g., John Giles Milhaven, *Toward a New Catholic Morality* (Garden City, NY: Doubleday and Co., Inc., 1970).

9. Fletcher, op. cit., p.75.

10. Fletcher, op. cit.. p. 22.

11. Soren Kierkegaard, *Fear and Trembling*, tr. by Walter Lowrie (Garden City, NY: Doubleday, 1954), p. 66.

12. Martin Buber, *Eclipse of God*, (New York: Harper, 1957), p. 118.

13. In Walter Kaufmann, ed., *Existentialism from Dostoevsky to Sartre*, (New York: World, 1957), pp.345-369.

14. Ibid, p.358.

15. William A. Spurrier sees validity in both natural law theory and in situation ethics, and thus attempts to formulate "a new synthesis." See his *Natural Law and the Ethics of Love*, (Philadelphia: Westminster Press, 1974).

16. Robinson, op. cit., p.15.

17. *Situation Ethics*. See also, "Love is the Only Measure," *Commonweal* (Jan. 14, 1966), pp. 427-432.

18. John A.T. Robinson, *Christians Morals Today*, (Philadelphia: Westminster Press, 1964), p. 23.

19. *Situation Ethics*, op. cit., p. 155. Also, Fletcher himself moved from a Christian theistic understanding of love to a non-theistic humanist understanding. See his "An Odyssey: Theology to Humanism," in *Religious Humanism*, XIII, No. 4 (Autumn, 1979).

20. Erich Fromm, *The Art of Loving*, (New York: Harper, 1956).

21. Ibid., p. 26.

22. Fletcher deals with this case in two works: *The Ethics of Genetic Control* (Garden City, NY, Anchor Books, 1974), pp. 134-35, and *Humanhood: Essays in Biomedical Ethics* (Buffalo, NY, Prometheus, 1979), p.133. My discussion is based on his chapter on "Abortion," pp. 132-39..

23. Fletcher comments on this case in *Situation Ethics*, pp.37.39.

24. For criticisms of Fletcher's theory of situation ethics from within the religious establishment, see *The Situation Ethics Debate*, ed. with an Introduction by Harvey Cox (Philadelphia: Westminister Press, 1968).

8.

Eugenics and Biotechnology

Herman J. Muller's View of the Scientific Future and Its Relevance to the Humanism of Today

Stephen P. Weldon

Man is a prober and meddler and in this, so long as he holds true to his own gifts, he will not stop. Undoubtedly what he will know and be able to do along biological lines only a few generations hence would seem like a science fiction dream to most of us of today. . . . I am convinced that, unless he shortsightedly destroys himself, as by means of radiation, he will remake himself. (Herman J. Muller, "Man's Place in Living Nature" *Humanist* 1957, pp. 101-2.)

Herman J. Muller, the sixth president of the American Humanist Association and one of the leading figures in 20th century genetics, was an unapologetic eugenicist, and his 1935 book *Out of the Night* makes an interesting case study in humanistic bioethics in the 20th century. That book represents something of an anomaly in the American humanist tradition in its blending of scientific futurism and ardent Marxism. In this respect, the young Muller stands much closer to left-wing British intellectuals than to his American colleagues. Muller eventually renounced communist ideology and Soviet politics. That early Marxism, however, continued to shape Muller's thinking about the relationship of science to society, and he maintained a strong commitment to the notion that scientists had a responsibility to humanity for the knowledge and the technologies that they made possible. Although I find Muller's ideas about eugenics and biotechnology fraught with problems, I think that some of his ideas about the relationship of science to the social sphere are worthwhile considering. Indeed, it is largely the inconsistent way that Muller treats eugenics in light of that social relationship that makes me critical of his thinking about biotechnology.

Muller's prominent place in the development of genetics provided him with an understanding of the possibilities of biotechnology, and he envisioned technical achievements in genetic manipulation that now seem prescient. Both the achievements and the defects of Muller's ethics are worthwhile considering. The questions that I pose at the end of the paper are meant to force humanists to consider how much they find Muller's views worthy of contemporary consideration. The intellectual and cultural transformations of the 20th century have exposed some of Muller's greatest weaknesses. A study of his work, however, reminds us of forgotten insights.

Herman J. Muller was born in 1890 in New York City and took his Ph.D. at Columbia, where he studied genetics in the *Drosophila* lab of T. H. Morgan. Morgan's lab was the premier research school in the country for studying the newly developed science of Mendelian genetics. During Muller's study there, Morgan and his students did path-breaking work in genetics and the chromosome theory of heredity. Muller left the lab to pursue research on genetic mutations and eventually received the Nobel Prize in physiology and medicine for his discovery that X-rays could cause mutations in genetic material.[1]

Muller was a political idealist throughout his life. He embraced Marxist politics in his early years and a realist-internationalist perspective in later years. In the late 1920s as a young faculty member at the University of Texas, he served as a secret faculty advisor to a communist front student group in Austin. He eventually moved to the Soviet Union, where he expected a politically progressive climate, and it was during this period that he published his main eugenics tract, *Out of the Night*, where he applied science to a thoroughly Marxist political agenda. Within only a few years, Muller became convinced that he had to leave Russia and Stalin's totalitarian regime, which had grown ever more hostile to his work. (Muller would eventually denounce the politically powerful Lysenko for his rejection of Western genetics and espousal of politically correct but scientifically questionable hereditarian ideas.)

Muller is still recognized for his scientific accomplishments in the field of genetics. Less remembered now, however, is Muller's eugenic thinking in which he provided a vision of the tight intertwining of scientific practice and social ideology. Eugenics has become a taboo subject. After the holocaust and its indelible mark on recent history, eugenics and the eugenic application of genetic selection have consistently met vehement opposition among people of most political and ideological persuasions. Previous to the abhorrent Nazi application of eugenic ideas, however, the idea that societies should attempt to control the human gene pool was not only broadly acceptable, it was openly promoted.

Eugenics originated from the ideas and work of Francis Galton in England at the end of the 19th century. He and Karl Pearson developed the notion that people could consciously direct the evolution of human populations by examining their biological traits and employing social controls that would encourage the propagation of good genes and discourage bad ones. The ideas gained such currency and popularity that they infiltrated into British and American politics. America's restrictive immigration policies of the early 20th century were a reflection of this. Eugenics even permeated popular culture, so much so that American fair-goers were able to participate in family gene competitions during the 1910s and 20s.

Eugenics was not all of a kind, however. The historian Daniel Kevles divides eugenics into two camps: the popular, conservative eugenic programs of the early 20th century, which he labels "mainline eugenics"; and the socially progressive ideology, "reform eugenics." Conservatives embraced a form of eugenics that sought to freeze the social order as it was, while progressives found in eugenics the possibilities for dramatic social transformation. Mainline eugenic policies tended to favor immigration restriction and efforts aimed at discouraging population increases among members of the lower classes, who were often seen to be mentally defective. Race purity and, especially in England, aristocratic privilege lay at the heart of mainline eugenics. Reform eugenicists—Muller among them—denounced such policies.

In their view, disease and fitness was not located in a specific class or race, rather it spanned the human species. Eugenics offered a method of ridding the gene pool of deleterious genes wherever they were found, without the same elitist ideological discrimination against whole classes of people.

Muller's major work on eugenics was a book entitled *Out of the Night*. It was published in 1935, but the seminal ideas originated in a student literary society lecture he had given as an undergraduate in 1910 at Columbia. Muller revised the paper for lectures at the University of Texas and the University of Chicago in 1925, but held off publication for several years. Not until 1935 did he find the time ripe to publish the book. Muller, then living in Russia, sent a copy to Stalin, thinking its advocacy of Marxist socialism would appeal there. Instead, it met hostility owing to its radical proposals regarding sexuality and reproduction. The book fell on deaf ears in the United States as well. Only in England did *Out of the Night* gain an enthusiastic following. Indeed, the combination of overt Marxism and scientific futurism found a ready audience among British left-wing intellectuals. In response to Muller's advocacy of sperm banks, for instance, George Bernard Shaw announced in his typical bombastic manner: "Think of all the ova I might have inseminated!!! And of all the women who could not have tolerated me in the house for a day, but would have liked some of my qualities for their children!!!"[2]

The basis behind Muller's eugenic program was a method whereby the human species could gain control of its evolutionary potential. It went much further than much eugenic thinking which sought only to prevent the deterioration of the gene pool. Muller explicitly wanted to improve the human race. For him, human biology was as malleable as clay. One need only master the techniques to manipulate the genetic material. Eventually, he maintained, eugenicists would attain such subtle control over the intricacies of human biology that human psychological characteristics like intelligence and altruism could be affected.

Muller, at the leading edge of modern genetics, knew quite well how the new genetic understanding of biology could create dramatic possibilities for reshaping human beings. Technologies such as artificial insemination could be made available, which would have major repercussions on how eugenics programs could work in the future. Old-style eugenics had been based in large part upon controlling and directing marriages, a very ordinary and conceptually simple plan. Artificial insemination, however, would make it possible to sever "the function of reproduction from the personal love-life of the individual." Men and women could marry for love but give birth to genetically superior children. Muller was ecstatic about the opportunities: "Only social inertia and popular ignorance now hold us back from putting [such practices] into effect (at least in a limited experimental way."[3] In future, Muller mused, even more extensive genetic control would be possible, including parthenogenesis, foster pregnancy, and direct manipulation of the genetic material itself.

A long-time science fiction fan, Muller speculated about the furthest reaches of technological control over human genetic material. Muller imagined even further that scientists would someday be able to engineer new kinds of people: "the invention of new characteristics, organs, and biological systems . . . will work out to further the interests, the happiness, the glory of the god-like beings whose meager foreshadowings we present ailing creatures are." Humans could create "a being beside which the mythical divinities of the past will seem more and more ridiculous, and which, setting its own marvelous inner powers against the brute Goliath of the suns and planets, challenges them to contest."[4] Later, he would claim that the current form of man was merely "a transitional phase" in the evolutionary scheme of things.[5] The very foundations of human understanding and behavior could be transformed through genetic change. His eugenic ideas were visionary and utopian.

The technology would become so advanced that it would transform the basic human functions of reproduction and childbirth. Muller, following the ideas of the British socialist J. B. S. Haldane, promoted a technology called "ectogenesis," whereby fetuses could be allowed to grow entirely outside of the womb. Women, he stated, had been restricted in society by their biology and their role in the bearing and raising of children. Ectogenesis and other reproductive technologies would liberate them almost entirely from what he considered to be a terrible burden.[6] Gone from eugenics were the traditional methods of selective marriage and sterilization. Biology and technology fused in a world that radically altered the nature of humanity.

When one looks at the end-product of Muller's vision, in fact, there seems to be nothing left of human beings in the traditional sense. Our bodies and minds have become objects of manipulation and the result of that genetic work is entirely foreign to the common understanding of the term *human*. Apart from obvious questions about ethical choices and political control, one must ask whether anything labeled *humanism* can be sustained under such a vision. The very definition of humanity gives way in such a world.

Yet, rather than focus on the foreign nature of Muller's vision, I want to look more closely at the specific social and political ideas that Muller adopted in 1935, compare that to his post-Marxist thinking on eugenics, and finally consider the ways that biotechnology has come to be employed today. Here, I think, lie some of the most troubling issues that humanists must deal with.

In *Out of the Night,* Muller made a clear argument about the relationship between science and the social world. The social environment, he argued, must be changed *prior to* the implementation of any eugenic program so that genetic change would be directed toward the proper political ends. In his view, this meant that the socialist revolution would have to take place first before any eugenic technology was employed. He was driven by an explicit Marxist ideology in which science serves the social order. Unless eugenic technologies were implemented after a socialist revolution, any eugenic technology could be used to strengthen the status quo and further the racism and elitism that already existed. The thing that Muller most lamented in *Out of the Night* was that eugenics had been too often put in the service of reactionary and racist ideologies. In Britain and America, capitalist ideology and aristocratic mentalities were directing most of the eugenic programs. Unless the emphasis on competitive capitalism shifted toward cooperation and equality, eugenics programs would be deleterious.

His Marxism reinforced his radical eugenic ideas as he considered what the proper relation should be between science and society. The society had to be in charge of the science, in his view. This did not, of course, mean that scientific progress would be hindered in any way. As should be clear by now, Muller was one of the most radical and utopian of scientific thinkers in the 20th century. Nonetheless, he understood that his science was to be employed for proper political ends and, hence, was to be directed by political institutions in the social sphere. His was not a laissez faire model of scientific advancement. The people had the right and the responsibility to control new scientific and technological ideas. This is Muller's singular insight that too often gets forgotten in debates over progress and science.

After his break with Marxism, Muller continued to maintain the necessity of social activism in the effort to control the applications of science. This ideal was manifest most clearly in Muller's efforts in the early Pugwash conferences that brought scientists together from Russia and the West to discuss nuclear proliferation and the dangers of nuclear war. The motivation behind these conferences was that the world's scientists had a responsibility to clarify the understanding of the implications of nuclear technology. The first conference was initiated by a joint statement from Albert Einstein and Bertrand Russell and signed by Muller and several other prominent scientists in the West. The attendees of the conference discussed the science and politics of nuclear weapons with an eye to finding ways to prevent a nuclear catastrophe. Muller's active promotion of these conferences, in spite of his complete distrust of Soviet intentions, highlights his embrace of internationalist politics and his commitment to a responsible use of science and technology.

It was this internationalist point of view and his progressive understanding of the relationship between science and society that brought him and the American humanists together. He was elected president of the AHA during the 1950s, and in 1963, Muller was elected Humanist of the Year. Those were the years that he was especially active in the Pugwash conferences.

Muller continued to espouse the development of human biotechnology. The *Humanist* published two important statements of his vision in the 1950s, in which he again articulated his extreme futuristic vision of human beings remaking themselves biologically into better physical, mental, and emotional organisms. Devoid of the Marxist rhetoric, these statements waxed romantically about the benefits of biotechnology and the grand vision of a future in which the human organism is remade.

I detect, however, an emptiness in these new visions where there had once been a thoughtful principle. In the young Muller's Marxist eugenics, it was mandatory that the social world change first before one could expect beneficent results to come of scientific advances. The result of an unregenerated society employing eugenics would be disastrous. Technology would be employed toward the wrong ends. Indeed, one would think that the Nazi atrocities committed in the name of eugenics would have been an immediate cautionary tale.

It is not that Muller entirely avoids cautioning us about the need for social change, for he continues to maintain the necessity of it, albeit with a different political slant. It is rather free inquiry that guides his social vision: "Man in the shackles of authoritarianism is incapable of such advances. Should he attempt them, his efforts would be misdirected and corrupting." It is only "when he is free to see things as they are" that he will be able to rightly direct human evolution toward greater heights.[7] There is a much greater sense of optimism and inevitability in Muller's new vision. *Out of the Night* maintained that a social struggle was necessary. The only struggle in Muller's later pieces is the effort to free people from "the shackles of authoritarianism." When people are thus freed, knowledge and understanding of science will enable them to utilize technology properly. Muller has become a laissez faire liberal in his outlook on eugenic/genetic engineering.

I'm not advocating a return to Marxist socialism as a prerequisite for social engineering like Muller did in 1935. Such a direction could only be authoritarian. Rather, I am pointing out that the Marxist point of view included within it a very important idea, namely that the social realm is ultimately where notions of responsible use of technology should be discussed and played out. It should not be left either to the scientists themselves—which is another type of elitism—or to optimistic ideas about how better knowledge and freer thought can yield good answers to these questions. Only thoughtful discussion and debate in the social sphere—informed by good scientific understanding, but not dictated by it—will yield what I believe to be a proper way of negotiating the difficult decisions that lie ahead in the realm of biotechnology and social engineering. In this respect, I'm much more cautious than ever Muller was. Not sharing with him his scientific utopianism, I am also more willing to give voice to criticisms of science and technology when they are necessary.

There are serious problems in asserting that any one vision will get the right answer. As Muller discovered, Marxism had great dangers, and Soviet Russia produced a society that was far from benign. The question that we must all face today, humanists and non-humanists alike, is what we are going to do with these new technologies that Muller so presciently prophesied. They are not fully upon us yet, but no one can think that they are just science fiction visions any longer. Artificial insemination and foster pregnancy are commonplace, in part due to Muller's promotion of sperm banks. In some ways the ethical difficulties that Muller spoke to have realized themselves. We now do live in an age in which semen is stored and people who have the money and ability can shop around for the qualities that they like in children. The technology has become capitalized, and the market will drive it, for good or for ill. This is eugenics in the broad sense of the term, but there is nothing in this vision any longer that matches the old-style state-sponsored eugenics programs. This new sort of eugenics brings with it new questions.

In addition, there are some ethical questions that Muller never asked, perhaps never could ask because of his broad vision. He couldn't see the individual trees for the bird's-eye view of the forest he chose to take. These close-up questions arise out of the particulars of the actual technologies as they exist in practice. What do we make of the enormous waste that occurs during embryonic implantation. We now have a vast store of frozen embryos that currently rest in limbo. What is their legal and ethical status? There are those in this country who would like to grant them the legal rights of American citizens, while there are others who would prefer to see them as prime research material. Or what about the children of foster pregnancy who want to know who their genetic parents are? Their social and psychological identity is tied to their biological make up. What legal and ethical rights should they have when they go searching for people who have never met them and perhaps never want to do so? Or, even more problematic, what are the ramifications of preventing the birth of children with genetic defects? Can we just leave this decision up to the parents, or to the mother alone? Is this a personal decision or a political one? How does it bear on what we declare to be human?

None of these questions are new or unique. We've all thought about them and discussed them. They are becoming platitudes. The point I want to make by reminding us of Muller, is to reaffirm Muller's conviction that those who advocate and are willing to promote technologies have a great responsibility to think through the careful and wise use of those technologies. Humanists, by and large, are unwilling to put great restrictions on science and scientific advances, seeing in science the kinds of salvation that Muller envisioned. In recent years, in the wake of postmodernism and the rise of religious fundamentalism, humanists have become ever more sensitive to any criticism of science and technology, fearing that they might be abetting a reaction against progress. As a result, they tend to be reluctant to make sharp criticisms when the need calls for it. Laissez faire seems to be the usual response: issue quiet injunctions against inappropriate use, but let science be.

Can we go beyond the platitudes, however, and genuinely seek realistic ways of thinking about biotechnology without being afraid that we will be "anti-science"? Perhaps there are times in which to be a humanist means to be anti-science. The ambiguities of biotechnology are legion and need to be fully explored. Humanists of the 21st century should learn from Muller's example, consider the strengths and weaknesses of his approach, and think about how historical events have reshaped the world that Muller spoke about in the early 20th century. Where do we go from here?

BIBLIOGRAPHY

Muller, Herman J. "Life." *Humanist* 15 (1955), pp. 249-261.

Muller, Herman J. "Man's Place in Living Nature." *Humanist* 17 (1957), pp. 2-13, 93-102.

Muller, Herman J. *Out of the Night: A Biologist's View of the Future* (New York: Vanguard Press, 1935).

Carelson, Elof Axel, *Genes, Radiation, and Society: The Life and Work of H. J. Muller* (Ithaca, N.Y.: Cornell University Press, 1985).

Kevles, Daniel, *In the Name of Eugenics: Genetics and the Uses of Human Heredity* (New York: Alfred Knopf, 1985)

NOTES

1. Biographical material about Muller from Carlson, *Genes, Radiation, and Society* (1981) and Kevles, *In the Name of Eugenics* (1985). There is also an interesting unpublished essay on Muller by Scott Mellon, "Mutation and Eugenics in the Thought of H. J. Muller" (Master's thesis, University of Wisconsin—Madison, 1989).

2. Quoted in Kevles, *In the Name of Eugenics* (1985), pp. 191-2.

3. Muller, *Out of the Night* (1935), p. 111.

4. Muller, *Out of the Night* (1935), p. 125.

5. Muller, "Life" (1955), p. 259.

6. Kevles, *In the Name of Eugenics,* pp. 184-186.

7. Muller, "Life," (1955), p. 261.

Putting Bioethics to Work

9.

Biology As a
Window into Therapy
Ethical and Practical Considerations[1]

Michael Werner

In our family we, like all parents, are continually asking our children why they persist in a behavior we've told them is clearly wrong. Of course asking a child "why" is really a pretty fruitless endeavor. In fact, why do any of us do the things we do?

Of course, human behavior, as a consequence of psychological inquiry and the theories that try to explain this behavior, is better understood today than ever before. But we still have far to go. We know that we are only at the beginning of a scientific understanding of how and why human beings behave as they do. Psychological therapies are rapidly evolving as they benefit from scientific psychology and from the empirical evidence developed in the other biological sciences. So if, as is sometimes said, the twentieth century was the "century of physics"—e.g. relativity, quantum mechanics, nuclear energy, space exploration—the twenty first century will probably be called the "century of biology." As with so many other modern developments, the new biology challenges traditional moralistic interpretations of behavior and, not least of all, challenges existing corrective therapies.

Our ideas about human behavior have undergone a long series of revisions. At one time, mental illness was, and still is in many quarters, seen as demonic possession, as sin, or as moral weakness. With the advent of Freud and modern psychology—e.g. behaviorist, developmental, gestalt—mental illness was interpreted in more secular terms. Although Freud aimed for a scientific basis for understanding behavioral and emotional problems, his overall approach was more clinical and even literary than scientific. Often brilliant in his insight into the human condition, he nevertheless failed to meet the model of inquiry shaped by the experimental sciences. But, he did pioneer the search for an empirical understanding and treatment of mental illness that persists to this day. To put it in overly simple terms perhaps, rather than blaming the person for what was now seen as an illness, we were encouraged to try to find both the etiology of and the treatment for what were now understood as patients and not as villains.

In this essay I will use the treatment of drug and alcohol abuse as an instructive example of how our understanding of behavior and our response to behavioral problems have evolved as a consequence of the new biology. Traditionally, alcohol abuse was viewed as a moral weakness. In a more puritanical America, it was interpreted as a religious failure; in other cultures as an ethical defect. It was Benjamin Rush, a physician who lived at the time of the American Revolution, who was one of the first to recognize that the persistent urges of alcoholism arose from factors other than immorality. Rush was the father of the assertion that alcoholism is a disease and therefore treatable.[1] However, the message was not heard. Instead, reformers in the nineteenth century created prison/asylums where addicts were kept in solitary confinement for a year or more. They lived in monastery-like bare cells in the hopes that contemplation of their sins would bring them to their moral senses. Instead, from what we now know of the effects of such conditions, it apparently drove many mad.

In the past century alcohol and drug treatment in the US has been dominated by Alcoholics Anonymous [AA] and similar twelve-step programs. So prevalent did this approach become that 93% of all treatment programs in the US use them today. AA emerged out of the Oxford Movement. Promoted by Bill Wilson, the AA model views users as spiritually flawed and in need of giving their life to some trans-natural power, to "God as you know him." In fact, of the twelve steps, only two actually reference alcohol. The twelve-step approach is premised on the notion that the alcoholic by himself or herself is powerless, i.e. no one ever entirely "recovers" from alcoholism. Emphasizing its moralistic and religious nature, the alcoholic, as part of the "program," must make moral inventories, and must make amends for past behavior. AA is a process based on the belief that alcoholism can only be dealt with by some form of god-belief, religious submission, and atonement.

There is scant scientific evidence that AA has much effect other than as a psychosocial support group. Its aim is control but not cure. However, its proponents maintain that it is the only way to sobriety.[2] By contrast, several studies have shown that 70-75% of addicts get over alcohol problems by themselves without AA or its like, or without other therapy.[3] Our current President, George W. Bush, is a prominent example. Similarly, the vast majority of those who quit tobacco use do so all by themselves. In fact, most people abandon their addictions when the perceived consequences come at too high a cost.

The AA/twelve-step model began as one where both the cause and addiction control were religiously based. Late in the 60s however, the medical model gained credence. It asserted that there was some organic defect that resulted in addiction. Ironically the twelve-step movement adopted this model although it was at odds with the religious defect model. One suspects that a less than worthy reason for adopting the medical model was that, as an official disease listed in DSM IV, treatment of alcoholics became eligible for insurance coverage. Thus the 28 day "miracle" treatment programs were established. Today the cognitive dissonance within AA is even more acute. The program asserts that the cause of alcoholism is biological but nevertheless insists that the response must be spiritual!

Meanwhile others saw the cause of alcoholism as primarily psychological. In other words, substance abuse was seen as a coping mechanism for dealing with psychic pain or trauma. In "reality therapy," which supports a pluralistic concept of psychological causality, the cognitive is dominant. As Albert Ellis and Emmet Velton write, "The most important causes of your addictions are your thoughts, attitudes, images, memories and other cognitions."[4] Philip Tate in *Giving Up Alcohol* says that there are two types of addiction both of which are psychological.[5] One arises from low frustration tolerance, the other from general emotional disturbance, Obviously, a therapy that is highly oriented towards a cognitive approach represents a radical shift from AA's way of dealing alcohol abuse. For AA, as we have seen, addiction is a religious issue where the individual is powerless; for cognitive therapies, alcohol abuse recovery is wholly a matter of "rational" choice. Addictions have both their cause and cure in our thinking.

An extreme example of this latter view is offered by Jeffery Schaller, a disciple of the psychiatrist, Thomas Szasz. He argues that addiction is a myth because addictive behavior can only be stopped by exercising free will. For Schaller, then, there is no organic component in causation since one's choice alone can be curative. Given this relationship between free will and behavior, Schaller sees the disease model as just wrong. "Drugs don't cause addiction. Nothing can addict a person," he says.[6] Therapy for Schaller involves dealing with life's existential problems and nurturing one's rational free will.

THE SCIENCE OF THE MIND

The polarization and the dissonance that can be seen in these various approaches to addiction are the effects of simplistic, dogmatic, secular and religious ideology. None of these succeed in arriving at a scientific understanding of human behavior. Strangely, they seem unaware of current biological research.

Recently there have been great advances in the science of the mind and we continue to uncover its inner workings. We now know that Cartesian mind/body dualism is simply false. The mind is more like a "committee" than a rationally functioning mentality embedded in a physical body. Marvin Minsky of MIT in his early work trying to duplicate the human mind with the use of computers found that the mind functions very much like a set of connected parallel computers processing many "algorithms" all at once. The mind does not process data in a linear fashion, but rather all at once using as it were many small "programs."[7] Decision-making is messier than linear rationality. The elements of many messages, all interacting, and some in deep conflict with each other, are contained in the process. Daniel Dennett uses the metaphor of "Brainstorms" to describe this highly complex process. While notions like the Freudian "subconscious" have been proven inadequate, we now believe that much of the process of decision-making is indeed hidden from us. As William James said, "A great many people think they are thinking when they are rearranging their prejudices."[8]

A multiple voice model of mind is being confirmed daily by research like that conducted by Antonio Damasio who is studying patients with damaged ventromedial prefrontal connections to the mid-brain (where emotions and basic survival skills reside). Damasio has found that patients exhibit a lack of moral center and an inability to decide amongst various alternatives for future action.[9] This occurs despite the fact that such patients can reason perfectly well in other contexts. It seems that, while our rational neocortex can weigh various cost-benefit options, the final decision may well be made by our more primitive Stone Age voices. It's as if this subconscious primal voice acts both as the final arbiter in close decisions and as leading moral salesperson in the committee of our minds.

Many other studies have found that the older midbrain was originally an organ for stimulating the organism to take action on important aspects of survival and reproduction. In lower animals that lack a rational reflective consciousness, behavior is motivated by endorphins such as dopamine that produced effects of elation; "feel good" chemicals if you will. We too are similarly affected. We don't think at 5:00 p.m., "Gee, I should eat or I will starve." Instead we get a gnawing, visceral urge that is dissipated after we eat and another endorphin, serotonin, tells us we are satiated. Endorphins motivate us to action in the absence of self-conscious reasoning. For those like human beings with self consciousness, they act as a competing powerful motivator. Addictive substances turn on our "feel good" chemicals such as dopamine.

In dealing with addiction and therapy, the contentious and unresolved nature/nurture debate has always been in the background. Now, the study of mind has illuminated the issues in that debate, the issues of choice and behavior, in a new way. Although many have criticized its science and particularly its perceived politics, it is nevertheless the case that evolutionary psychology does enlighten us about our motivations and our decision making behaviors. In *The Blank Slate*, Steven Pinker points out that in the twentieth century, three ideas have emerged from the study of cognitive evolution. One is that the mind is grounded in the physical world, i.e. there is no "Ghost in the Machine." The brain operates like a computer for processing information, computation and feedback. Secondly, Pinker claims that the mind is not a "blank slate." Indeed, there is a universal aspect to human nature and human cognition which would be impossible if environment was all. Lastly, Pinker claims that an infinite range of behavior can be generated by finite combinatorial programs in the mind.

By contrast, a good deal of human psychology in the past hundred or so years was based on a behaviorist model. The mind was thought of as a "blank slate" on which the environment writes. Given that view, the mind can be socially engineered to produce the desired behavior. Trying to get the environment "right" has led to unproductive and often repressive social and political schemes. Conversely AA holds, as it were, to a "Ghost in the Machine" model. The mind is something that can only be influenced by a power outside of nature and therefore the person, the minded being, is inaccessible to scientifically based methods. In a different way, Cartesian mind/body dualists including some in cognitive therapies see the mind as something apart from the physical universe and so as equally inaccessible or understandable by biological methods.

Many worry that in adopting a biological view of psychological behavior we must opt for some form of reductive determinism. On this view, free will and choice in our decisions becomes only a myth. Of course, nothing can be further from the truth. For the modern view of the mind that we have been exploring, human behavior is seen as a complex interaction of all of the brain's processes. A decision regarding our addictive behavior or any decision in fact is based on a confluence of voices in that meeting of the committee of our mind.

A pluralistic bio-psycho-social model called for here provides understanding for both the causes and treatments of addictions. It abandons the "all or none" attitude of most models and sees all the forces acting upon us and that we act upon as formative, but not determinative. It sees all these forces as proclivities, tendencies, and drives with none of them providing the complete answer. All of the forces are interactive and not really separated from each other.

Addictive behavior may initially be motivated by one thing and then may continue for entirely different reasons. For example, a person may start using alcohol and drugs as a method of self medication for depression. Alternatively he or she can start using to get a pleasurable high. Another person may start using as a way to be accepted by his or her peer group. As time and use progresses, the urge to continue may increase as a result of these and other interactive forces including organic changes in the brain as a result of use itself. Thus, we can see by PET and NMR scans that the "old brain" lights up due to the effects of endorphins. Synapses build pathways into our cognitive centers and even our neocortex physically changes over time.[11] As the brain changes physically during the addictive process, the distortive effects of addiction are more and more rationalized, protected and disguised. At first, the addictive behavior may result for primarily psychological reasons. Over time, however, the physiological aspects seem to become more dominant. While we do not yet know exactly how all this physical change in the brain takes place, the body of knowledge is increasing rapidly. It is probable that a more complete understanding of how this change occurs is within reach in a few years given the new analytical tools at our disposal.

Some would say that these biological changes prove that addiction is an organic disease like any other. But, if alcohol and drug addiction is a disease then it is the only disease we know of thus far that can be cured by psychological power alone. It turns out that while addiction is the result of a complex interaction of many factors, the cure can only be found in the power of our own brain. Our choice in the matter can indeed trump all the other factors. On the other hand choosing in the context of addiction is not a simple process and its must be reinforced by other processes as well. Choice is not easily available when in the depths of severe addiction. The urge to continue an addictive behavior may well highjack the self-control centers of our brain. The urge to continue using may overwhelm the message of sobriety in the battle between the committee members of the mind. Sobriety, in other words, does not always win. For example, recovery can be as low as 6% in methamphetamine addicts.[12] Generally, recovery is a process whereby the messages that give us permission to use are thwarted, sidestepped and destroyed while the messages to maintain sobriety are enhanced. One of the greatest tools for recovery is to reveal and dispute the lies to ourselves that rationalize our behavior and to replace them with messages of long term sober health. That is why "cognitive behavioral "methods are so successful.[13] It is why I have been a strong advocate of SMART Recovery™ and cognitive behavioral methods in general including Rational Emotive Behavioral Therapy (REBT). These methods, when enlightened and evolved by the new science of the mind generally do not fall into the trap of seeing the cause and cure of addictions as only psychological.

THERAPY, BIOLOGY, AND FREE WILL

The process of recovery from addictive behavior raises a larger question: how free are we to control our behavior? Thus far, we have used addictions as a tough example, a hard case. In the larger context, the issues are usually framed as a question of determinism. Is the person responsible for his or her ethical and, in this specific case, is the addict responsible for his or her addictive behavior. Are we in possession of free will? Traditionally, ethics has been an intellectual process. From what we have learned and are learning about the brain and about addiction, we would argue that ethics is best understood as a biological process intertwined with cognitive and cultural processes. In other words, the road to ethical wisdom lies through biology and the social sciences.

According to John Dewey, many philosophical problems are not so much solved as bypassed. Mind-body dualism and the nature-nurture argument have been rendered irrelevant. So too, it would seems with the ways in which the problem of free will has been formulated and discussed in the past.

Free will is not something we have, it is something we practice. As Steven Pinker points out in his book, *The Blank Slate*, "The fear [of determinism] is that an understanding of human nature seems to eat away at the notion of personal responsibility." What seems to have gone wrong "is a confusion of explanation with exculpation." To "explain behavior is not to exonerate the behavior."[14]

Daniel Dennett in *Freedom Evolves* has also tackled this subject of biological determinism. He writes, "if we accept Darwin's 'strange inversion of reasoning' we can build all the way up to the best and deepest human thought on questions of morality and meaning, ethics and freedom."[15] His definition of freedom as, "the capacity to achieve what is of value in a range of circumstances" is "as good a short definition of freedom as could be."[16]

Holding people accountable for their behavior serves as a deterrent to irresponsible behavior. One need not claim that we have "free will" or get into the metaphysical dilemmas of determinism in order to understand that behavior has many sources. One of them is cognitive, knowing we will be held responsible for our actions. Gaining sobriety is an example of decision-making in the light of this knowledge. People decide to change their behavior when the negative consequences become too high to tolerate. Of course, there are always some people whom no consequences including death will deter. Still, the vast majority of people can be motivated to change their behavior when they choose not to accept the negative consequences any longer.

So it is with any ethical decision. We will act ethically when there are significant rewards for good ethical behavior and there are significant costs for poor ethical behavior. Addiction, its processes, outcomes, causes and cures provide an ethical and not just a therapeutic model. Remember that no one wants to quit using alcohol and drugs after becoming addicted. It is only when the consequences of continued using are too high that behavior changes. Of course, "too" high is not a fixed and rigid norm. It is the individual who decides if the long term cost of using is more expensive than the short term pleasure of getting high.

A biological interpretation of decision-making requires a more nuanced, interpretation of human behavior. For example, Tourette's syndrome seems to interfere with areas of the brain dealing with control functions. Ask a "Touretter" to stop a tic and in many cases he or she can do so for short periods of time. But then he or she tics twice as much later on. It's not a matter of free will vs. determinism, but of choosing within a spectrum of possible controls over behavior. There are "Touretters" who have absolutely no control over some tics, but others who have a good degree of control. Traditional either/or notions of free will miss the point in such circumstances.

So it is with addictions. The urge to continue using after becoming severely addicted is extremely powerful and cognitive control is very difficult. What those of us in the addictions field have found though, is that when someone uses a phrase like "I can't stop," what he or she really means is "I really don't think I can stop because the urge to use is so high and there is intense pain when I don't act on that urge." This is particularly evident in the later stages of addiction. The addict takes a few drinks or "hits" and it's "off to the races." There are biochemical and physiological reasons why people go on binges other than just psychological and behavioral ones. The serotonin levels which tell our body we have had enough are severely reduced.[17] To this, we add the fact that a predictable addictive pattern of behavior has been built up. As the same time, the addict is cognitively impaired. Given this complex mix of factors, It is easy to see how the "committee" of the mind is stacked in favor of what Jack Trimpey calls the "Addictive Voice"[18]

Still, everyone seems to have control over that first use or "hit." So, remembering helps. Recovery involves the belief that one should stop, that one can stop, and that one has some tools to help in the process. The most important tool is a commitment to sobriety.

It is the same with any moral behavior. Commitment to a plan of action in the face of a set of circumstances can make all the difference. Ethical theory is useless unless we are personally committed to moral behavior. Jonathan Glover in *Humanity: A Moral History of the Twentieth Century* describes those who kept their moral standards in the face of socially accepted genocide. They did so not because of some ethical abstraction, but due to empathetic commitments to human compassion, connection, and decency.[19] Our empathic commitments can certainly still be influenced, manipulated, urged and subtly directed by the many voices in our committee of the mind. Accepting the responsibility for our choices and commitment to a plan of ethical action are not certain to result in changing behavior. They are not the only factors in making actual moral decisions. Still, just as it has been found in addiction treatment, it is long term commitment to a plan of action that is successful in changing behavior. It also appears to be true that long term commitment to empathetic, moral behavior is crucial to long term ethical behavior.

If we separate explanation from exculpation as Pinker urges us to do, we find that acceptance of responsibility is in fact a requirement for choice. Without acceptance, we are, as it were, putting our minds on "automatic control." The committee of the mind can operate without conscious reflectivity on the consequences of our behavior. For addictions, this failure to rationally reflect means listening to the messages to continue use and not disputing the developing excuses for our addictive behavior. The consequences are continued addictive behavior.

Our minds are fantastically intricate machines adapted for thinking, deciding and adjusting our behavior so that we can survive and reproduce. Decision-making, be it about addiction or any moral issue, is an intricate dance of many colliding, interacting and competing factors. The future of ethics and therapy will both involve a greater appreciation of the biological considerations. The empirical knowledge we gain in a bioethical approach will help us as we more fully understand how human decisions are made and which factors are most effective in achieving the decisions we want.

NOTES

1. William White, *Slaying the Dragon: The History of Addiction Treatment in America*, (Chestnut Health Systems Publication, Bloomington IL, 1998), p. 1.

2. Ibid., p. 127.

3. George Valliant, *The Natural History of Alcoholism Revisited*, (Harvard University Press, Harvard MA, 1995).

4. Albert Ellis and Emmett Velton, *Rational Steps to Quitting Alcohol*, (Barricade Books Inc., Fort Lee NJ, 1992), p. 67.

5. Philip Tate, *Alcohol: How to Give it up and be Glad You Did*, (Sharp Press, Tucson AZ, 1997), p. 46.

6. Jeffrey Schaller, *Addiction is a Choice*, (Open Court, LaSalle IL, 2000), p. 119.

7. Marvin Minsky, *The Society of the Mind*, (Touchstone Books, New York, NY, 1985)

8. George Seldes, *The Great Thoughts*, (Ballantine Books, New York, NY, 1985), p. 205.

9. Thomas Damasio, *Decartes Error*, (Putnam Books, New York, NY, 1994)

10. Steven Pinker, *The Blank Slate*, (Viking Press, New York, NY, 2002), p. 179.

11. Ronald Ruden, *The Craving Brain*, (Harper and Collins, New York, NY, 1997).

12. Lowinson, Ruiz, Millman, and Langrod, *Substance Abuse* 3rd Edition, (Williams and Wilkins, Baltimore MD, 1997)

13. Reid Hester and William Miller, *Handbook of Alcohol treatment Approaches, 3rd edition*, (Allyn and Bacon, New York, NY, 2002)

14. Ibid., p. 10.

15. Daniel Dennett, *Freedom Evolves*, Viking Press, New York, NY, 2003), p. 307.

16. Carlin Romano, *Dennett's 'Freedom' Reconciles Free Will with Determinism*, (Philadelphia Inquirer, March, 9, 2003).

17. Ibid., p. 11.

18. Jack Trimpey, *The Small Book*, (Lotus Press, Lotus CA, 1989), p. 63.

19. Jonathan Glover, *Humanity: A Moral History of the Twentieth Century*, (Yale University Press, New Haven CT, 2001).

10.

Bioethics, Contraception, and Business

Vern L. Bullough

We tend to think of bio-ethical dilemmas in terms of others, not something we ourselves are involved in. Sometimes, however, we are personally involved not only as victims but as conscious or unconscious advocates of ethical violations. To illustrate this quandary, I have used "population planning," as an issue in which I have been directly involved not only through my writings but through lectures and through personal decisions. I think the battle over the ability to plan pregnancies led to one of the greatest victories in the twentieth century and has radically changed the role and place of women in our society. This is a tremendous accomplishment but sometimes in our drive to change things, we might well have given lip service to what could be regarded as poor ethical decisions or did not openly protest ethical violations.

The ability to plan when and if a woman would become pregnant changed the way women looked and thought about themselves. It was the decisive factor in the second wave of the feminist revolution which occurred in the 1960s and 1970s and remains a strong factor in the lives of all of us today. The decision whether to use a contraceptive, however, is more than a decision for each woman to decide by herself or in collaboration with her male partner. Among other things it involves scientific research, the ability to disseminate information, and the availability of a product. It also involves a mind set about its desirability. This is where individuals often found themselves in compromising situations. Businesses in various forms—manufacturing, retail and wholesale distribution, and advertising—also faced ethical issues which they often failed and still fail to understand. Too often the main consideration has been either a profit one or a willingness to overlook ethical lapses by individuals and corporations for what was seen as a greater good, namely the desirability of family planning and population control. Often it is only after a disaster of one sort or another has occurred that ethical issues rise to the surface. Other times it is simply that a more expensive form of contraception is pushed when a less expensive one which is still useful is dropped. Sometimes it is an image problem in that a company does not want to be associated with the topic even though they have an effective product. Complicating the issue is the real fear that many people feel about the dangers of overpopulation and the desirability that significant sections of society sees in limiting population growth.

The greed is easy to document. A good illustration of it has to do with the commercialization of intrauterine devices (IUDs). Such devices have an ancient history but they remained problematic because off the possibility of infection occurring is a real danger when a foreign device is inserted into the uterus. This problem potentially was solved by the improvement in sterile techniques and materials and especially by the discovery of antibiotics which could overcome any infection. IUDs came to public attention in the U.S. in 1962 as a result of an international conference sponsored by the Population Council and almost immediately caught on. They were inexpensive to make, could easily be inserted by a physician, and could be removed (again by a physician) if pregnancy was desired. Lippes Loop (invented by Jack Lippes) cost less than a penny to make and was very profitable both to the company who manufactured them and to the physicians who inserted it. So profitable that other rival inserts with slightly different design (many of which were untested) were almost immediately put on the market by competing manufacturers seeking to benefit from this bonanza.

Such a success often results in bioethical dilemmas since new products, only with slight changes, are rushed on to the market without adequate testing, something not then required by the FDA. The A.H. Robins Company bought the rights to the Dalkon shield developed by a physician, Hugh Davis, and his business associates who had sold it to Robins without doing any real testing although they claimed to have done so. Neither was it tested by Robins. It was made of the same material as the loop but it had several sharp pointed edges to distinguish it from its rival. Unfortunately these edges tended to increase the dangers of infection. By 1976 some seventeen deaths had been attributed to it, yet the company ignored the complaints and discounted the deaths until 1980 when it finally advised physicians using it to remove it. The subsequent lawsuits bankrupted the Robins Company but they also raised the liability rates for other manufacturers, so much so that the IUDs on the American market more or less disappeared.

The Lippes loop, however, continued to be made and its use was widespread around the world because it was so inexpensive, could be easily inserted, and had a long history of success. There are three bioethical issues here: the first was the A.H. Robbins company and its search for a product to compete; the second was the physician Hugh Davis who performed few tests and more or less ignored negative results; but the third was the decision of other manufacturers to withdraw their IUD products from the market in the U.S., resulting in the loss of one of the most effective and inexpensive of birth control devices.

When the IUD reappeared on the American market in the 1990's, it did so in new forms which cost more (and gave greater profit) such as the Copper T which as of this writing is the most effective contraceptive on the market although IUDs are still not very widely used in the U.S. One of the difficulties with the IUD from a manufacturer's point of view is that once inserted it does not have to removed for a long period of time. This meant that there was no renewable business as with the pill, the condom, spermicides, or other birth control devices which kept the customer buying regularly. The manufacturers of IUDs coped with this by turning to a new type of IUD, using them as a means to deliver hormones, the same hormones available in the pill. This meant this kind of IUD had to be replaced periodically every year (although some have a longer interval) and both the physician and the company could contemplate a regular replacement policy. As the case of the Lippes Loop and the Copper T illustrates, economic issues, such as need for regular replacement, are more important than effectiveness. Money and the profit motive have to be considered a major variable in influencing bioethical decisions.

But profit itself is not necessarily enough. A good example is what happened to the polyurethane sponge impregnated with a spermicide, 250 million of which under the name of *Today*, were sold between 1983 and 1995. It was not as effective as the IUD but it had a long history going back to at least Biblical times, and its commercial form which included a spermicide was widely used by younger women with an effective rate equivalent to that of the diaphragm. The safety and effectiveness of the sponge was certified by the FDA and as its sales would indicate was popular with large groups of women. Nonetheless when the company which manufactured it, now called Wyeth, found it needed to upgrade the New Jersey factory where it was made, they decided that it would be better business practices to simply quit manufacturing it and concentrate on the pill which they also made. The panic which this caused was immortalized in a 1995 episode of the television series "Seinfeld" where Elaine, hearing of the decision to stop its manufacture, ran around New York seeking to find a supply of sponges, the method of birth control she favored. When she located a whole case at a pharmacy, she purchased all of them. She then proceeded to stretch her supply by giving would-be bed mates a test to determine whether they were "spongeworthy" before having sex with them. As of this writing, a startup company, Allendale Pharmaceuticals, which ultimately purchased the rights from Wyeth, has begun to again put them on the market. Like the condom, the sponge does not require a prescription, an important issue for many would-be users. Spermicides themselves can be sold over the counter and are easily available in cremes, jells, vaginal inserts, and other forms, but they are not as heavily advertised as other forms of contraception. Both the *Today* sponge (which is treated with a spermicide) and condom when used with a spermicide are very effective. Unfortunately most barrier methods such as the diaphragm or cap require a prescription to purchase in the U.S. and this requires a visit to a physician, making these barrier methods somewhat expensive, at least initially.

Oral contraceptives, better known as the pill, were approved by the FDA in 1960, and they were an instant money maker. One militant feminist, hostile to the drug industry, described them as giving the manufacturer a guaranteed customer for thirty or forty years of her life for preventing a disease (pregnancy) which could not be cured and which always demanded a replenishment with only a few interruptions for a planned pregnancy. The ethical issues involved here are the real lack of testing of the pill and the more or less repressed representation of the early experimental data. The public was tremendously receptive and more or less unquestioning while the manufacturer had a monopoly which it did not care to share.

The first pill, *Enovid*, included 10 mg of progestin and 150 micrograms of estrogen and had considerable side effects for many women. Yet these side effects were downplayed. In fact in the first widespread experimental use of them in Puerto Rico large numbers of women became so nauseated and ill that they dropped out of the program. The medical community more or less ignored such side effects until militant feminists staged demonstrations against them beginning in late 1969. The result was a reformulated pill which are either low progestin pills (without estrogen) or combination pills which include 1 mg or less of progestin and 35 to 50 mcg of estrogen, a rather radical change from the original pill, and indicative of the importance of the protest movement in changing the prescription. Most of the pills today are triphasic, with greater or lesser amount of the hormones depending on the time of the month. Note it was only demonstrations by women that forced a reconsideration of the original formula. Threats and demonstrations sometimes seem to bring about greater consideration of bioethical issues.

Money issues are complicated by other aspects of doing business. The company which held the patents to *Mifepristone* or *RU 486*, an abortifacient developed in France, fearful of a boycott of its other drug products by anti–abortionists, abandoned its potential market in the United States. No American company was immediately willing to step into the gap. A special company had to be set up by devoted advocates to manufacture the pill and do the necessary tests for FDA approval. It is now marketed in the U.S. by Danco Laboratories, a company independent of any major drugs manufacturer which was set up to do the job. When taken with another drug, usually misoprostol, a prostaglandin pill to make sure the accumulated material is expelled by the uterus, it causes an abortion. Misoprostol is marketed in the United States as *Cytotec* where it has been approved as a treatment for gastric ulcers. So worried that its drug might be associated with abortion, Searle, the manufacturer, issued a national letter to health providers stating that *Cytotec* was not approved for the induction of labor or abortion. This, despite the fact, that it is used widely to induce early labor in obstetrical patients. In a sense, Searle is technically correct. The drug has not been approved for inducement of labor, simply because its manufacturer has not asked for approval in this capacity, although the company surely knows what it is used for. Is this ethical? When it comes to controversial drugs, bioethical issues seem to have many dimensions, including denial and cover ups.

The same ambivalence about ethical responsibility is apparent in the unwillingness of manufacturers of oral contraceptives to indicate that they can be used as emergency contraceptives to terminate a pregnancy before the fertilized egg is implanted. Technically they have not been approved for this purpose by the FDA but both physicians and the FDA well know how effective the pills can be if used in this way. They are fearful of a backlash.

This use of oral contraceptives is not complicated at all. When a woman realizes or believes she might have engaged in an unprotected act of intercourse because she had neglected taking an oral contraceptive or only irregularly used them, an unusual dose of pills will prevent pregnancy. One doesn't need to take them every day but only in such emergencies. For pills with 50 micrograms of estrogen and 0.5 milligrams of progestin, such as *Ovral*, a total of four tablets should be taken in divided doses, an initial two, and the other two twelve hours later. The series must start within 72 hours of unprotected intercourse, and it is desirable to start it within 24 hours. Other oral pills with lower doses of hormones such as *Nordette, Triphasil, Levien, Tri-Levin, Lo-Ovral*, require dosages of eight pills, four at each interval instead of two. The mini pills require even greater dosages, e.g. *Ovrete*, require 40 pills. Should not this information be indicated in the usual literature accompany the pill? Shouldn't this possibility be widely publicized? Isn't this a bioethical issue?

What has happened in the contraceptive field is a search for new ways of administering the hormones associated with the pill without having to actually take a pill. The IUD, as indicated, now can do this but so can inserts and injectables. Norplant inserts under the skin in the arm can, depending on the number in the pack, last for three to five years after which they have to be removed. Biodegradable capsules or pellets inserted into the uterus are effective for 18 to 24 months before they begin to disintegrate. Injectables such as *Depo-Provera* or *Megastrom* (both are technically depot medroxyprogesterone acetate) can be given, depending on formulation, every day, once a week, or every twelve weeks. Pills have also been developed to be taken over longer periods with the result that a woman if she is willing or desirous of a need to menstruate not more than four times a year, can change her cycle to do so. Vaginal rings, which encircle the cervix, release hormones which make the cervical mucus impregnable to sperm. They can be worn for three weeks, removed for one, and then reinserted. There are also hormonal patches. Take your choice. Most of the various alternatives, however, require a visit to a physician and periodic replacements, all money in the bank for the business and the physician.

Sterilization is always a possibility with tubal ligature in either males or females, and it is possible to undo it if the attempt is done early on. Again, however, this requires medical intervention. Unfortunately, for many years, in states such as California, sterilization was done regularly on poverty patients and on mentally deficient or mentally disabled patients, almost always without their consent. The governor of California in 2003 issued a public apology for these sterilizations done in the name of eugenics, a procedure which many of our Humanist pioneers advocated. Today there is a new method called *Essure*, a tiny coil that can be inserted into the fallopian tubes by a catheter and which allows tissue to grow around it blocking the tube. It is easily reversible in the first few years after it is done. Sterilization by any means should require much discussion between the medical provider and the client and if he or she has a partner, with the partner. Unfortunately, this is not always the case.

One of the worst abuses I have noticed in my research took place in China where female sterilization was done by chemical scarring of the fallopian tube. Women who underwent it after having had one child were promised all kinds of rewards including guaranteed college admission to their first child and significant (for China) monetary reimbursement. The procedure could be done in outpatient clinics and involved almost an assembly line procedure taking five minutes or so with one patient being prepared for the procedure, a second undergoing the procedure on one fallopian tube and a third on the other fallopian tube. It entailed inserting a catheter through the cervix into each fallopian tube duct and injecting an acid formulation. The acid would then result in severe scarring, thereby blocking the tube.

The procedure was done without anesthetic. Physicians assured me it was painless but watching the patients afterwards clutching their stomachs and bending over in pain indicated to me that this was not quite true. Some seemed to be shock. I wrote up my experience for a medical journal and also talked about it to others indicating my belief that some anesthetic should be used. Surprisingly, or perhaps not so surprisingly, many of the people felt that the goal of cutting population growth in China should overcome my concern for a woman who only had a comparatively brief period of patient pain. They believed that the dangers of over population were so great that criticism such as mine would lead to a curtailment of sterilization in China. Unfortunately, many of those who criticized me were nonbelievers, but hopefully not humanists.

Quite clearly the whole area of birth control and population planning raises all kinds of bioethical issues. Humanists have long been concerned with rapid population growth and the threat it poses to the survival of humanity. It is a legitimate concern. It is tied into a number of other issues which some of our free thought forbears did not handle very well. Many of the believers in and advocates of eugenics were affiliated with the free thought movement. They felt it was important to sterilize the feeble minded and the handicapped, and some even went so far as to urge such drastic measures against people whose only problem was that they lived in poverty. Some physicians almost routinely did hysterectomies on poor women who had had many previous pregnancies and who they felt should not have more. Unfortunately they often did not ask these women for permission. These were common practices often done without malice. Today we recoil, or at least some of us do, in horror at what went on.

Yet, there are no simple self-evident answers. Do we want those who are barely able to take care of themselves and whose intelligence is that of a six or eight year old to have a child? Or, to make someone else pregnant? With injectables and implants, there are alternatives to sterilization. But how much say should the patient have in making such decisions? How much should the dangers of overpopulation weigh when decisions are made about a person who cannot cope in society without major help is likely to become pregnant or make someone pregnant? How much should manufacturers reveal about their products? Should the whole business of contraception be a money making one in which it is financial reward not public need which drives the way decisions are made?

In the abortion area, the bioethical issues are compounded. Here we have vested interests, including the U.S. government, giving out misinformation. For example, the Center for Disease Control, under the influence of President Bush recently included a statement that abortion might be a causal factor in cervical cancer. This, in spite of the fact that only one paper found a possibility that it might be a causal factor something that more than 200 other research papers deny. In fact, that one paper has been seriously undermined by follow-up studies. Should deliberately false propaganda be a bioethical issue? The Bush government has cut U.S. monetary support of birth control programs around the world. Should the personal belief of a powerful official undermine basic human needs in a country or in a world which disagrees with this opinion?

Quite clearly, contraception and birth control pose serious and difficult ethical dilemmas. This emphasizes the all-encompassing nature of bioethical issues and the difficulty of dealing with them. The enemy, if there is one, is too often ourselves.

SOURCES

This paper is based essentially upon my own published research including Vern L. Bullough and Bonnie Bullough, *Contraception: A Guide to Birth Control Methods* (Buffalo: Prometheus Books, revised, 1997), and Vern L. Bullough, *Encyclopedia of Birth Control* (Santa Barbara: ABC Clio, 2000).

11.

Bioethics and Justice
Economics, Care, and Conflict

Carmela Epright

In this paper I will introduce two separate, yet equally important, approaches to moral decision making in a clinical setting. In order to do so, I discuss the facts of a clinical case that took place almost a decade ago. It is a very famous case; one with which you may already be familiar. Let me begin by explaining why I chose to discuss a 10-year-old case study—I do so for several reasons: first and foremost I want to discuss the *Lakeberg* case, because it is certainly the most dramatic, if not the most significant, case that I have had the privilege of watching unfold as a relative "insider." It was also the *first* clinical case that I witnessed from this perspective. For these reasons, this case had a profound effect upon the ways in which I conceive of the clinical aspects of medical ethics. Secondly, as will become clear in the discussion that follows, this case raises a number of questions. I use it to discuss the role of patient/surrogate autonomy in clinical decision-making and to address the ways in which these sorts of decisions contribute to or detract from the interests of social justice.

These concerns are very sharply defined and delineated here—indeed, the conflicting principles of autonomy and justice are more clearly juxtaposed in this case, than in any other that I have seen or read about. Finally, I discuss the *Lakeberg* dilemma, because the error in reasoning that operates here—and it will become clear that I do believe that an error in reasoning was at work—continues to be reproduced in other clinical cases. So, I elected to go back 10 years to revisit a case that then dominated discussions in health care in order to determine whether we the bioethics community learned any lessons from it. Sadly, I must conclude that although bioethics has come a very long way over the course of the last decade, the concern that I will raise here is, I believe, as troubling today as it was in 1993.

LAKEBERG CASE

Early in 1993, 16 weeks into what had seemed a normal pregnancy; Reitha (Joey) Lakeberg was informed by her doctors at Loyola Medical Center (LMC) that she was carrying seriously disabled, conjoined twins. Although they considered other options, the Lakeberg's ultimately decided to carry the pregnancy to term—they told their physicians that they wanted "to hold out hope that the doctors were wrong, to believe that the babies would be born healthy."

Unfortunately, no such miracle occurred. On June 29, 1993 twin girls, Angela and Amy Lakeberg were born even more significantly disabled than doctors had predicted. The infants were joined breast to belly; they shared a liver and a heart. While it would be possible to split the liver, the heart was severely abnormal—it had 6 chambers instead of 4, and a hole in the top that received blood from the lungs into the wrong chamber. Without surgical intervention this heart would not support the life of either twin, and no amount of surgical intervention would permit the heart to support both twins.

Following the birth of the Lakeberg twins, the staff at LMC recommended that the infants be kept comfortable and allowed to live out what was expected to be a very short natural life. Separation was not recommended. "We sort of pleaded with the parents to take the babies off the ventilator," said the neonatologist who tended the twins. "I suggested that we feed them and keep them as comfortable as possible, that we put them 'in God's hands,' so to speak."

However, the *Lakeberg* parents wanted to pursue further treatment. The staff at LMC refused to perform the surgery because they believed it to be medically futile. However, at the parents' insistence, they did put the Lakeberg's in touch with surgeons at Children's Hospital in Philadelphia where the parents persuaded doctors to perform a separation.

It was known from the beginning that both girls could not survive the surgery. The sacrifice of Amy would be necessary in order to provide Angela with a 25% chance of surviving the separation procedure and a 1-5% chance of "long-term survival," (5+ years).

As expected, Amy died during the surgery. Over the next 8 months Angela sustained numerous additional medical procedures.[1] She survived with the assistance of a negative pressure ventilator (similar to an iron lung). She was able to be held by her mother one time, but was never able to leave the hospital. Finally, she died 20 days shy of her first birthday.

The costs associated with the twins' care were not disclosed by either hospital, but they are thought to exceed 1.5 million dollars at the Philadelphia hospital alone. The Lakeberg family carried no private health insurance, and Medicaid refused to reimburse either hospital for most of the expenses associated with the twin's care because the case was deemed "medically futile."

"If someone is going to ration care because of money, it is not going to be us," said one of the attending physicians at Children's. However, one might argue that rationing *did in fact take place* in the *Lakeberg* case. LMC and Children's Hospital turn away hundreds of poor, uninsured families each year—as do all hospitals. Thus, the decision to spend approximately two million dollars on two patients with such a poor prognosis could be read as a decision *not* to allocate those resources to other patients. One commentator noted, "We have kids who have no immunizations and mothers who have never received pre-natal care—who have just given birth and don't know how to feed their babies." We know that immunizations and pre-natal care save lives, and we know that these treatments are inexpensive. Thus it is clear that this money could have been used to save the lives or insure the health of tens—if not *hundreds* of sick children with a far better chance for long-term survival.

I will begin my analysis by calling attention to the fact that this case violates one of my central pedagogical rules. When I teach clinical ethics to undergraduates, medical students, and residents I generally try to steer clear of "headline grabbing cases." I stress that ethical issues need not garner nationwide attention to introduce profound philosophical concerns. I urge my students to look for moral moments in *quieter*, less obvious examples—in the everyday interactions friends and lovers, parents and children, patients and physicians, family members and members of the healthcare team.

This particular case was in every respect a "headline grabber." It captured the attention of the nation from the twin's birth in June of 1993, until Angela's death on June 9, 1994. Pictures of the *Lakeberg* twins would ultimately appear on the covers of *Time* and *Newsweek*, and the front page of the *New York Times*. For several weeks this case framed the national debate over health care policy and cost containment. It spurred discussion over such concepts as "quality of life," "health care rationing," and "medical futility." It forced the general public to reflect upon the sort of distinctions generally considered only by philosophers: between what we *can do* and what *we should do;* between preserving life and prolonging suffering, between killing, and letting die.

In the interests of full disclosure I should tell you that I have a personal connection to and interest in this case. In the winter of 1993, I was a graduate student in bioethics participating in clinical ethics rotations at Loyola Medical Center. By chance, I met Reitha "Joey" Lakeberg the day that she learned that she was carrying conjoined twins. It was my first day in this (or any other hospital) as anything other than a patient. It was this case that served as my introduction to the clinical aspects of medical ethics. Needless to say, I discovered very quickly that ethical questions in a clinical context are not merely theoretical constructs, and they are quite unlike the scenarios that we ethics professors dream up to inspire our students. They involve real persons who stand to suffer or flourish under the weight of our investigations and decisions.

Obviously the controversy surrounding the Lakeberg twins raises more issues than I can address in this paper. For example, this case convincingly shows that we should be deeply concerned about the role played by the media in shaping newsworthy cases; it raises important questions about surrogate decision making for persons who have never been competent; and it demonstrates the different and often contradictory obligations that healthcare workers have to their patients and to science.

I employ this case here to raise some specific points about the tension between two different approaches to bioethics, "clinical medical ethics" and "organizational medical ethics." What I will refer to as "clinical medical ethics" or "case-based ethics" attempts to alleviate the concerns of individual patients and caregivers in particular cases. It strives, for example, to uphold patient autonomy, to distribute beneficence while avoiding paternalism. "Organizational medical ethics" on the other hand, focuses less upon individuals than upon *institutions*. For example, it attempts to institute equitable and just policies for health care delivery in light of scarce resources, and multiple, compelling claims to health services.

Most of the decisions in the *Lakeberg* case were made under a clinical ethics model. For this reason, this case is frequently cited as an example of just how difficult it is to exercise institutional decisions in particular cases, to do what is referred to as "rationing at the bedside." As several commentators have noted, such rationing decisions are dubious specifically because it is not immediately obvious that denying care to say, the Lakeberg twins would necessarily have meant that the funds would be allocated to more "worthwhile" causes. Would the funds have been used for prenatal care or vaccinations? We do not know; and we have no mechanism in place to allow us to answer such a question. In the absence of national, publicly accountable standards of care and clearly defined rules for health care expenditures, rationing certainly could and probably does occur—insofar as some people get a great deal of high quality care and others get little or none—yet these rationing decisions can hardly be said to be based on any sort of reasonable calculus. Given the current arbitrary and capricious nature of our system, the relationship between one expenditure and another is difficult, if not impossible to trace.

One of the assumptions that I will make in this paper is that if rational standards are to be created—and I hold out hope that they can and will be—institutions will have to make difficult decisions *in advance* concerning the sort of procedures towards which they will and will not allocate scarce resources. At the same time, they cannot be rigid and must also provide ample space for re-examination of such policies, in exceptional cases. Given this assumption, I hope to begin to bridge the gap between the concerns of clinical and institutional ethics—to ask whether the lessons learned in clinical practice can be used to govern institutional policy with respect to health care justice; and conversely, whether a concern for justice can help us re-think our approach to individual autonomy.

CLINICAL ETHICS AND THE LAKEBERG TWINS: AUTONOMY VS. JUSTICE

Eleven days after the birth of Angela and Amy Lakeberg the twins remained ventilator dependant. All available medical evidence suggested that even if the twins were separated, the surviving twin would never live free of respirator dependency. The ethics consultation team at Loyola Medical Center convened to review the case argued against separation surgery on three grounds: First, a careful calculation of the likely medical outcome indicated that given the configuration of the heart, only one twin could survive the separation surgery, and there was virtually no chance that the surviving twin would leave the hospital alive. This view was supported by the fact that all previous separation procedures of this sort (where twins had been joined breast to belly) had resulted in the death of both twins. In the most medically successful instance, the separated twin lived three months past the separation procedure, after having sustained numerous additional procedures and surgeries.

Secondly, the team cited the general principle of "non-maleficence." In this case "doing no harm" meant refusing to bring about the death of Amy, an unavoidable consequence of the separation procedure. The ethics team noted that they may have been inclined to override this principle if acting to cause the death of Amy—who was certain to die in the near future, whether or not the surgery was performed—would spare the life of the Angela. However, as noted above, there was no medical justification for believing that the separation procedure that surely would kill Amy would save Angela.

Finally, the team noted that the interests of social justice demanded that the hospital's resources (time, energy, technology, and professional expertise) should be devoted to cases that promised better outcomes. Thus, the ethics committee and physicians at Loyola had reached the same conclusion; everyone involved in the care of the

Lakeberg twins at this hospital had reached. They all advised against the separation surgery.

Meanwhile, the Lakeberg family remained committed to doing everything possible to save the life of at least one of the twins. They appealed to the hospital for a second review of the case, and were disappointed when a new board (composed of entirely different members) reached the same conclusions. At the parent's insistence, LMC put the Lakebergs in touch with Children's Hospital in Philadelphia. Surgeons at Children's agreed with Loyola's staff about the grim medical outlook for even one twin, and asked the *Lakeberg*'s to reconsider their bid for surgery. But when the parents continued to press, the staff at Children's relented and agreed to attempt the surgery. "We take the position that the parents have the right to choose for their children," says Dr. James O'Neill Jr., lead surgeon and spokesman for the Philadelphia team.

This one, brief statement, let me repeat it, *"We take the position that parents have a right to choose for their children,"* cuts to the core of the very different assessment of values underlying the conclusions drawn by Loyola and Children's in this case. Both institutions agreed that the separation surgery was certain to kill Amy and was highly unlikely to preserve Angela's life;[2] yet the staff at Children's chose to operate after the staff at Loyola had recommended that the surgery not be performed.

By way of explaining their discordant positions, each institution pointed to a different ranking of core values. As outlined above, the ethics committee at Loyola placed such principles as non-maleficence and justice above patient, or in this case, parental autonomy; but the staff at Children's was convinced that upholding the autonomy of the Lakeberg parents superceded all other moral obligations in this case. Moreover, in various published accounts, Children's rejected the notion that it was appropriate even to consider questions of justice or to raise concerns about resource distribution in the context of the *Lakeberg* case.

This is a fairly standard criticism against exercising the principle of justice in a clinical context. Taking justice seriously as a principle requires that one consider the interests of so-called "invisible others," such as future or hypothetical patients whom we have good reason for believing could, do, or will exist, but nevertheless are not immediately before us. Some contend, as did the staff at Children's, that it is inappropriate to speculate on future patients or to try to promote the welfare of such invisible others when one is faced with actual patients in a given case. From this perspective, it was the responsibility of the health care team to consider the welfare of the Lakeberg twins, not to speculate about how benefiting them might harm other, future patients.

I would agree, that it was not in fact the responsibility of the health care team to do such calculations. Their responsibility was to advocate the best course of treatment for the patient in front of them. In fact, this conflict gets to the heart of my central claim. In my view, it was the responsibility of the *institution* to set limits on what the physicians could or could not agree to provide to these patients and their family. While physicians and other clinical staff must do what is best for their particular patient(s), someone must "mind the store." The institution must think about the interests of all current and future patients. It must set guidelines for clinical practice so that there are available funds for future patients. Hence my distinction between clinical and institutional ethics.

For reasons that I hope will become clear (and which do not include alumni loyalty!), I heartily endorse the conclusions reached by Loyola's two ethics committees in the *Lakeberg* case. Yet I must also note that I find their ranking of values unusual, given my experience discussing ethics cases in a clinical context. I have a number of problems with expecting the values expressed by what bioethicists commonly refer to as the "Georgetown mantra,"[3]: i.e. autonomy, beneficence, non-maleficence, and justice, to serve as the basis for all discussions in healthcare ethics. Nevertheless, I will discuss these principles here, because they so obviously framed the debate over the *Lakeberg* case at both Children's and Loyola.

Although these four are intended to serve as *prima facie* principles—that is, principles that apply, all things being equal, and which are to be "ranked" only with respect to the *context* of a given situation—they are rarely exercised in this manner. If autonomy, beneficence, non-maleficence, and justice are intended to be of equal importance, then it must be said that in practice autonomy "is more equal" than the others. For better or for worse, the principle: "respect individual autonomy" has transcended the realm of philosophical discourse and has entered the vocabulary of medical decision-making.

In many respects reverence for individual autonomy is a vast improvement over past practices. No one would seriously argue that health care services were better distributed under the all too familiar "Father Knows Best" model: where the physician with *his* superior medical knowledge, was thought to be best suited to make all decisions for the sick (and thereby presumed incompetent) patient and the distressed family. Emphasizing the notion that patients and families may have projects and values that differ from the physician's and that individuals are uniquely qualified to decide what is in their own best interest, is one of the outstanding contributions that bioethics has made to medical practice.

Moreover, the principle of autonomy has allowed us to develop such necessary legal doctrines as informed consent; and concern for individual autonomy continues to deepen our understanding of mental competency whatever its current limits. Permitting patients and their surrogates to be self-determining, to make their own choices according to their own criteria is essential to recognizing the inherent worth of each individual. To deny such choice, is implicitly to claim that someone else knows what is best for them. This, as Kant aptly notes, serves to treat the person as less than a rational being, and to undermine his or her dignity.

PROBLEMS WITH AUTONOMY

But what exactly does it mean to uphold individual autonomy? The response given by Children's Hospital in the *Lakeberg* case, "we take the position that the parents have the right to choose for their children," implies that if the institution wanted to uphold the autonomy of the Lakebergs, the healthcare team was *required* to do *whatever* the parents requested. Even if the act in question were medically contraindicated, would result in harm to another, and might undermine the interests of justice, they were still required to obey.

Upholding autonomy in a situation where I have more information than the person in question clearly requires more than "doing whatever the person in question demands that I do." It requires providing that person with information so that he or she can make an *informed choice*—so that he or she will not be coerced by ignorance as it were to act on insufficient or inaccurate assumptions. I cannot say for certain that the Lakeberg parents were coerced by inaccurate information when they pushed for the separation of their daughters. When, however, he was informed that his daughter Angela had approximately a 1% chance of long term survival, Kenneth Lakeberg compared this to winning the lottery. As he said in words that became famous, "people win the lottery every day, why can't we?" For her part, Joey Lakeberg said that she couldn't live with herself if she had not done everything to preserve the life of her child. Neither seemed to understand that even if Angela survived, she would endure countless medical procedures and would be significantly disabled and respirator dependent. Nor did they acknowledge the impact that this decision was likely to have on Angela's 5-year-old sibling, their own relationship, or the long-term interests of their family.

In the end Amy and Angela were not the only members of the Lakeberg family to suffer. Two months after the separation procedure, financial pressures resulted in the family's eviction from their home. A fund set up to help the family was used by Kenneth Lakeberg to purchase cocaine and other narcotics—he ultimately became addicted. In interviews he attributed his inability to succeed in treatment programs to the stress brought on by his daughters' case. While his daughter was still hospitalized in Philadelphia, Kenneth ended up in prison on charges for robbery, assault and disorderly conduct. Following the death of Angela, Joey and Kenneth Lakeberg separated and their then six-year-old child was taken into the custody of Indiana social services.

I mention these disastrous consequences not to condemn the Lakeberg family, but because such outcomes are not unusual for families contending with significantly ill, disabled, and dependent children. Clearly such a high profile case was likely to generate additional stress. The institution should have been aware of the data suggesting that any family with a seriously ill child will likely suffer stress-related adverse consequences that could compromise the parents' marriage or threaten the well being of other siblings.

Did Children's Hospital uphold the autonomy of the Lakeberg parents by agreeing to perform the surgery, despite the likely dismal outcome of this case? I don't quite know how to answer that question.

Obviously, respecting someone's autonomy does not guarantee to make that person happy or to bring about a desired outcome. And, of course, a person can autonomously choose to act in ways that do not serve his or her own long-term interests. Nevertheless, I would feel much less concerned about the fact that this situation *literally destroyed this family*—if I believed that Joey and Kenneth Lakeberg had fully understood the medical implications of their decision to separate their daughters and if I was sure that someone had discussed with them the familial and social risks associated with caring for a significantly disabled child.

If the overriding desire to uphold autonomy at all costs were an uncommon approach to moral reasoning in a clinical setting, I could ignore the decision made in the *Lakeberg* case. I would "chalk it up" so to speak, to the exceptional, stressful circumstances of a media-saturated case. However, in the ten years since I watched the *Lakeberg* case unfold, I have witnessed scores of clinical cases that rely on this precise form of moral reasoning. The imperative for action becomes "do whatever the patient or surrogate demands."

I recently discussed the *Lakeberg* case with a group of physicians and clinical ethicists—and every single person in the room had a recent example of this sort of case. A neonatologist claimed that similar reasoning operates whenever an infant is born at the gestational age of 22 weeks. A mere three weeks of additional gestation would increase such an infant's odds of survival by approximately 70%. But, at 22 weeks, even exorbitant, extensive, and expensive intervention results in a survival rate of approximately 5%. Moreover, those children who survive such interventions will be significantly disabled, and will require a lifetime of therapy and support. In cases like those my neonatologist friend reports, his institution urges him to respect the family's autonomy—to follow the lead of the grief-stricken, confused parents, without intervention, that is, without providing direction or argument.

Others in our discussion provided examples of patients who, having been declared dead by all neurological criteria, were left to languish on life support for months or even years at the insistence of family members who "hope for a miracle." In each of these cases, the institution relinquished responsibility for these decisions, preferring instead to defer to the "autonomy" of the patient or family.

In such an atmosphere, it is difficult to ask other questions, to raise concerns about such principles as beneficence or non-maleficence. It is nearly impossible to raise broader more complex concerns about the effect that these sorts of decisions have on the distribution of health care resources.

Let me be clear, I am not suggesting that institutions return to the "bad old days' of paternalism, this time with the institution instead of the physician behaving as the parent. I do, however, wish to demonstrate that preferencing one principle—in this case, autonomy—at the expense of all other considerations undermines other, equally justifiable moral concerns and as I have shown, actually serves to distort the preferenced principle itself.

I realize that there is some danger in over-emphasizing an institutionally based approach. Institutions, like physicians, do not always know what is best for individual patients or their family members—indeed a patient's most deeply held beliefs or highest values may be completely different from the values held by the institution just as they may differ dramatically from the values held by his or her physician. Thus, an institutionally based approach will surely be inadequate in some cases. This means that institutionally based policies will need to be informed and supplemented by a continued commitment to clinical ethics.

It is my concern for and commitment to clinical ethics that forces me to call for institutions to take clinical based decisions seriously enough to embody some of what we have learned through clinical practice into policy decisions. If we have learned anything in the 10 years since the *Lakeberg* debacle it is that autonomy is a rich, complicated, and deeply problematic notion. Truly upholding autonomy means thinking carefully about the ways in which it is related to other concerns like a desire to do good and avoid harm, as well as desire to promote health care justice. This insight is so vital that it ought not be left for physicians to come to or not come to in particular cases while they struggle with clinical concerns, the needs of their patients and their obligations to those patient's family members. In other words, physicians are ill equipped to distribute justice at the bedside.

Distributing justice should be the responsibility of institutions, which ideally, are keeping track of their own cases and the cases faced by other institutions. It is the job of organizational ethics to institute policies to prevent us from repeating the sort of errors perpetrated in the *Lakeberg* case.

NOTES

1. Children's Hospital of Philadelphia and the Lakeburg family have declined to reveal how many invasive procedures Angela sustained, although both acknowledge that multiple procedures were indeed necessary.

2. Although O'Neill had argued against the claim that Angela's chance of surviving was a mere 1%, when pressed to explain her chances of survival he admitted, "if there is long-term survival, it would be unique."

3. Beauchamp TL, Childress JF. *Principles of Biomedical Ethics*, 4th edition. New York: Oxford University Press, 1994.

12.

Rationing Is Not
a Four Letter Word[1]

Howard B. Radest

The owl of Minerva spreads its wings only with the falling of the
dusk."
      [*Hegel's Philosophy Of Right*, T.M.Knox, Oxford, 1976, p.13]

THE SILENCE IS DEAFENING

With 45 million uninsured and alone among developed
nations, the United States fails to serve the health care
needs of all of its people. As it were, 80% of us benefit, in part at least,
at the expense of the remaining 20%. This inequality is, among other
things, an instance of implicit rationing and crude utilitarianism. Our
total health care budget in excess of a trillion dollars, i.e. $1.6 trillion
in 2003 and going up to $2 trillion by 2010 of sooner.[2] If this amount
were divided over the whole US population, health care would indeed
be universal. Unfortunately, however, given the structure of our medical care it would be even more inadequate than it already is. In other
words, we ration by interest, wealth, and power and yes, even by class.
But, as demographics change—the ratio of older to younger people,
the ratio of employed to unemployed, partially employed, and

retired—and as the economy recovers from "irrational exuberance," per capita resources decrease. Already we see this happening as the states try, unsuccessfully, to make up for funding limits in the federal budget. We cover less and less even for the fortunate 80% and we serve fewer and fewer people. Thus, the ranks of the "privileged" are shrinking and bound to shrink further. Nationally, health care serves— but barely—only because it is decreasingly national. We "cover" those whom we choose to cover, i.e. those who enjoy economic security and effective political voice.

No one would admit willingly to denying care to their fellow human beings. Good intention is everywhere the cynic might say. However, good intention is obviously not enough practically or morally. "To will the end is to will the means," as Kant put it. In this, both Christian and pragmatist, religionist and secularist agree. In other words, "by their fruits shall ye know them." Yet, we disconnect means and ends when it suits us. We escape responsibility by describing attempts to join purpose with practice as unworkable, as socialistic, as too costly or what have you. Subjective integrity—the comforts of a clear conscience, "I care, I really do"—is thus betrayed by the facts on the ground.

An ambiguous moral climate develops. For the majority of us, the present situation is or seems to be advantageous. So we have an interest in keeping things the way they are although we do not acknowledge that self-interest is at work. Of course, we allow minor adjustments or proposals for minor adjustments in order to satisfy the noisier outcries and correct the more outrageous—or the more vis- ible—violations of human decency. Consider the misguided "drug benefit" for Medicare recipients enacted into law in December 2003. As written, and apart from those provisions that clearly benefit corpo- rate interests and disadvantage government, the amount of actual sup- port for prescription drug costs is strikingly inadequate. For example, the amount of co-payment from the recipient in addition to an annual participation fee estimated at $420 (deductible plus insurance pre- miums), ranges from $438 of the first $1000 to $1500 of $3000 to $3500 of $5000 in expenditure by an individual for his or her pre- scription drugs each year. It is only after the $5000 mark is reached that anything like genuine help is given. Even then, a $10,000 annual drug bill will call for $3845 payment by the recipient, i.e. about 40%

of annual costs.[3] When added to other costs of care and other co-payments, e.g. for acute care treatment in a hospital, the program will obviously be of use only to those with middle to upper-middle class retirement incomes or first-rate insurance policies or to the very poor.

At the same time, we plead poverty while enjoying privilege. The situation is thus marked by moral illusion, an illusion which we genuinely believe is real. For example, typical discussions with physicians and policy makers seek to reassure us that even those not insured will get medical care somehow.[4] Resistance to change is therefore a function of consciousness, ideology, status, and interest. The "free market" provides us with moral cover since all of us "know" that the free market with its "invisible hand" is the most efficient and effective means of doing things. With Candide, we are convinced that "all is for the best in this best of all possible worlds" if not now then certainly in the "long-run." Of course, as Lord Keynes remarked, "In the long run you're dead!"

Our genuine charitable impulses are corrupted by our traditional Puritanism. Thus, the "undeserving" poor—unwed mothers, AIDS patients, drug addicts, alcoholics, welfare cheats, the homeless, the unemployed, the non-compliant, etc. The list of the unworthy grows—latterly, the immigrant illegal or even legal. Their moral failure—i.e. their sin—produces their economic and social failure. The responsibility, then is not ours but theirs! Indeed, shades of social Darwinism, it does morality a disservice to sustain the "undeserving" in their sinfulness and sloth by acts of charity.

This ambiguous moral climate is further reinforced by social myth: life is sacred, we say and human dignity is priceless. It is therefor just unthinkable that people like us would not do the right thing. Dramatic instances—the child in the well, the so-called "principle of rescue"—confirm us in our virtue. Thereby, we hide from the fact that we make life and death decisions all the time and often for economic reasons. We think to objectify these by calling them "cost/benefit" decisions, invoking a useable statistics and confusing it with determinism. We establish priorities, make judgments about what can and cannot be done for and to whom. All this, of course, in a condition of anonymity—the wondrous comfort of population statistics—and a denial of moral agency. The mythic pose assures us that we are not an unconcerned people.

In this context, I turn to my theme and, inspired by Spinoza's *Ethics* which has always been a favorite of mine, to lay it out as a series of axioms, propositions, and commentaries.

AXIOM 1—FINITUDE

Even if the issues of access were put to bed [i.e. issues of a just distribution], there would remain a second issue: availability [i.e. issues of a just allocation].

Commentary:

a) I think it was Emerson who wrote, "A man's reach should exceed his grasp or what's a heaven for." Sadly and ironically, in our time Emerson's glorious optimism must yield to the demands of a different reality. Short of heaven and living in our finite world, scarcity must sooner or later be confessed as humanity's fate. To be sure, scarcity is a moving target. What is in short supply today was simply a luxury [an option for a few] yesterday. Scarcity is thus both a natural fact and a social and psychological construct.

b) However, even if our voracious consumerism were to be tamed by stoicism, it remains our lot to face difficult—often seemingly impossible—choices where to assure the benefit of some we must assure the suffering of others, often helpless others like children and single parent families. Scarcity does not permit the illusion that with a change here or there or even with a radical shift in our habits, we will somehow ensure that everyone benefits and that no one suffers.

c) Thus, rationing under conditions of justice, i.e. under conditions of equitably shared burdens, entails that those who have will have less, sometimes, as in health care, significantly less, and that those who have not will have more but often, as in health care, not as much as they need. Little wonder then that the "haves" find ways to evade the issue. Meanwhile, the "have-nots"—who inhabit the same moral culture as the "haves"—cannot find their own voices. They too are caught in and believe the myth of the deserving poor. They too are convinced by the soothing tones of Harry and Louise, two characters in a very effective TV advertisement that helped to kill President Clinton's health care reform program in 1993. Those of us who have worked in the field know that nothing is so viciously judgmental as the poor about the poor. They too blame themselves. They too believe that all life is sacred. Little wonder that "rationing" hardly appears anywhere in political or social discourse..

AXIOM 2—ONTOLOGY

Infinite desires are incoherent in a finite world.

Commentary:

a) It is tempting to distinguish "wants" and "needs." Thus, John Rawls "primary" needs or Abraham Maslow's "hierarchy of needs." Then, consistent with our traditions, we can ease our conscience by restricting the rigors of rationing to the control of desire. The problem, however, is that self-interest shapes the argument, i.e. my "need" is your "desire." Consider in this regard the often-heard complaint that welfare clients have "big" TV sets, drive "fancy" cars, are lazy, etc. Of course most of us complainers have never lived on food stamps or welfare checks, in slum housing or on the streets. Further, in the market culture that is generated by a market economy, wants are converted very rapidly into needs and sometimes appropriately. Hence, the power that produces the sentence: "I have to have that!"

b) More generally, a notion of indefinitely increasing expectations—a culture of desires—subverts the distinction between needs and wants. By implication, efforts to determine a "minimum standard of health care" which should be available to all falls afoul of a culture that is unfriendly to minima, e.g. to Norman Daniels' normal functions, Amartya Sen's capabilities.

In order to unfold these considerations, I suggest the following propositions essential to establishing health-care rationing in a democratic society.

PROPOSITION I: ETHICS AND POLITICS
Legitimacy—both actual and visible—is an absolute requirement of any rationing scheme.

Commentary:

a) Democracy uniquely faces the dilemmas of rationing. Other political forms do not. The latter may make justice claims in order to stabilize political power, e.g. Bismarck's Germany, Castro's Cuba, or for public relations purposes and self-congratulation. In democratic states, however, inclusion and participation are necessary conditions of legitimacy. Obviously, these terms are easy to define but difficult to realize. For example, in a diverse and pluralist society with many sources of authority, culture, and value, inclusion will require empirical as well as normative judgments. "One size cannot fit all" any more than "one person, one vote" can assure more than formal citizenship participation. After all, even dictatorships hold elections.

b) To be legitimate is to be transparent. Following Kant again, that which is done in secret is morally suspect on the face of it. So legitimacy requires public procedures and public access to information. This should come as no surprise and is consistent with our constitutional tradition—open courts of law, due process, etc—despite lapses like Alien and Sedition Acts, Attorney General Palmer's raids after WWI, Japanese internment during WWII, McCarthyism in the 1950s, and the present Attorney General's "patriot act" in the light of 9/11.

PROPOSITION II: FAIRNESS
Impartiality must be and be seen to be uncompromised.

Commentary:

a) Legitimacy requires fair procedures, i.e. procedures which are open to all with understandable and publicly available rules, accepted by the polity, announced in advance, and applied impartially. But fair procedures while necessary are not sufficient. For example, it is possible that all of those in a similar condition could be treated similarly (a standard definition of justice) and yet the result could be morally wrong. Two historic examples come to mind. A fair procedural rule might be: only and all those with a certain minimum of real property are eligible to vote. Another might be: only and all first born sons shall inherit the family's property. Thus, too, the market-place can well be defended as procedurally fair, i.e. anyone who can afford some good can have that good.

b) Fairness requires attention to substantive outcomes, i.e. in some sense a morally acceptable end-state, that which is to be achieved by rationing, must be describable, attainable, approvable, and applicable. Outcomes are clearly historical and empirical. As culture evolves so do moral values, e.g. the changing meanings of "inclusion" as religion, status, ethnicity, race, or gender cease to be grounds of exclusion.

c) Both fair procedures and fair outcomes presume that "objectivity" is attainable particularly in making life/death choices and quality of life decisions. But objectivity is not self-defining. Hence the struggle to find methods of translating incommensurables into some kind of common currency or common measurements, e.g. QALYs—quality adjusted life years; DALEs—disability adjusted life expectation. But this adds further complications. The meanings of "quality of life" and "disability" depend on notions of acceptable and unacceptable life-styles, social values, the varied norms of communal cultures, etc. Reductionist strategies—e.g. pay attention only to "relevant" biomedical/clinical matters like diagnosis, treatment, and prognosis begs the question. To be sure, organ transplant policy may work in limited instances but cannot succeed as a model in global rationing schemes. Even in such limited instances, value judgments must enter. For example, is family support entirely a clinical requirement for approving an organ transplant or is it also a reflection of both middle-class values and social investment priorities.

d) In clinical and other health care decisions, intimacy and subjectivity cannot be evaded. It is, of course, arguable that they really cannot be evaded anywhere but that is another story. Ultimately, there are no "strangers at the bedside." A sense of common destiny pervades experience—there but for the grace of God—or good luck—go I. We do exhibit our capacities for empathy and sympathy. Indeed, absent these, and health care probably does not succeed as well as it could. The quest for valid uses of detachment, impartiality, and objectivity thus becomes even more problematic. That, no doubt, motivates recent efforts to develop moral narratives, to embed the clinical case in personal biography in order genuinely to acknowledge the primacy of the "patient." In this context, the distinction between "advocacy" and "interest" need to be specified.

e) Rationing entails democratic coercion. Both libertarianism and fascism fail the requirements of justice. Rationing may expect us, as John Hardwig wrote, to consider a "duty to die." Less dramatically, rationing may expect us to surrender beneficial treatments after a certain status is reached. Daniel Callahan, by way of example, suggests "age" as one such status. Ultimately, rationing requires shared deprivation. Further, rationing doesn't work unless its provisions are distributed over the population. The "free rider" sooner or later destroys any rationing system, i.e. subverts fairness, challenges legitimacy, etc. Exemptions and loopholes as with our tax system distribute burdens unfairly, i.e. the problem of an equitable graduated tax, and so subverts any rationing scheme.

f) Coercion does not necessarily entail the surrender of autonomy. We tend to treat autonomy and liberty as synonyms. But, while they are within the same conceptual neighborhood, they are not the same thing. Following Kant, yet again, autonomy means giving law to oneself. But to give law to oneself does not require that one has to create the law to which allegiance is given. Autonomy is also exercised when I make a law of a society, a community, a family or what have you, my own and freely do so. So, in exercising autonomy, I impose constraints on myself and even, in some instances, against my personal desires and interests, i.e. Ulysses contracts where I instruct others in advance of an event to prevent me from doing something which at a later moment I insist that I want to do. Thus, the democratic enigma: as it were, as Rousseau maintains in *The Social Contract*, I coerce myself.

PROPOSITION III: INTELLIGIBILITY
Rationing must be understandable and reasonable.

Commentary:

a) It is all too easy for expert knowledge to become manipulative and exploitive. Paradoxically, the more "scientifically based," the greater the possibility and the temptation of obfuscation and even duplicity. This is compounded by the fact that health-care traditions and practices are still attached to the mystique of authority and the mysteries of esoteric knowledge. Scientific and evidence-based medicine do not necessarily dispense with these traditions. But for rationing to succeed, its terms must be understandable and informed consent not merely pro-forma. Hence, rationing entails a serious threat to the ways in which caregiving has been understood and interpreted even under modern conditions and despite a modern commitment to patient autonomy. Rationing, then, poses a global and not just a specialized intellectual and pedagogical challenge.

b) Rationing requires semantic directness, i.e. a language that reveals rather than hides. By contrast, the current language of priorities, allocations, and cost/benefit only seems to be purely technical—the province of expertise—and only seems to be morally neutral. Analysis, however, reveals that these terms are never entirely descriptive. Allocation requires judgment about what is and is not to be included in the territory to be allocated and this judgment entails moral values both explicit and hidden. For example, consider the problem of the military draft, the values that do or do not authorize exemptions for women, students, married persons, conscientious objectors; consider, too, the use of the draft as a covert method for shaping the civilian workforce, etc. Similarly cost/benefit must answer to the questions: whose cost and whose benefit; what counts as a cost and what counts as a benefit? In answering such questions, special pleading, which is often unrecognized as special pleading even by those doing it, is typical rather than rare.

PROPOSITION IV: SOCIABILITY
The notion of the general welfare must be reinvigorated.

Commentary:

a) Born in the Enlightenment, the American republic embraced and embraces a radically individualist culture. While it is true that we are also "joiners" as DeTocqueville thought, it is as true that the values which lead to us to join are individual enhancements of one sort or another. Typically, we join for any number of personal ends from salvation to dating skills. While this generates a rich and varied culture and offers the benefits of choice, it also gives rise to a devaluation of community and of society. Thus, the "general welfare" as in public health is neglected except in crisis. Communal values are attenuated. For example, except for dramatic communal conflicts like those over prayer in the schools, "gay" marriage, or abortion, community too is understood as self-serving or self-enriching. Under these conditions, recognizing and acting as a member of community for the sake of community is hardly available. The common good is hardly real to us. The exceptions make this clear: e.g. religious communities like the Amish or the Hasidim.

b) For most of us, "social contract" seems a usable alternative. But "contract" calls for "give and get." Sociable behavior is rooted in an exchange of goods and preferences. It is the market translated into politics. It is, then, no accident that a market culture and not just a market economy evolves. Under contractual conditions, only fair exchanges are legitimate, i.e. I cannot, as Robert Nozick insists, legitimately be forced to surrender properly gained advantages or goods without appropriate payment. To command surrender of a good without a reciprocal reward is morally dubious. Debates about "takings" and libertarian views of taxation rest on the notion that self-interest is a moral claim. Only voluntary acts, acts of charity that do not require exchange are morally permissible. The state may encourage but cannot command such acts, e.g. tax exemptions for charitable contributions. But even here we notice that charity is in this way socially legitimated as another instance of "give and get." Thus "social contract" both presumes and reinforces individualism.

c) Neither individualism nor social contract can provide an effective democratic basis for rationing. Recall that the "haves" are to surrender a good while the "have-nots" are to receive a good. The former experience a cost; the latter a benefit. Under such a condition, fair exchange is not possible although there may be promises of postponed exchanges or what is called "enlightened self-interest," e.g. a healthier society benefits everyone; future generations will thank you. But contracts with non-existent future agents are not enforceable, i.e. those agents could not have signed the contract nor could they have designated a surrogate; their interests are unknowable, and in any event their interests cannot have been specified by them.

d) Fortunately, neither individualism nor social contract accurately reflects experience although our myth speaks to us as if they do. Persons are by nature sociable; human development is essentially communal. Exchange values may characterize certain but by no means all relationships. Compare, if you will, an auto-dealership with a family. Indeed, communal as against functional relations are highly valued, i.e. the values of intimacy, love, caring, acceptance, etc. Genuine regret at the absence of community is typical. Neither the hermit nor the functionary serves as a model. We disdain the "bureaucrat" often unjustly and as often for symbolic reasons. Without quite realizing it, we are responding to the moral limitations of contract.

e) For rationing to succeed then, a different realism and a different mythos is required. Connectedness, individuation as distinct from individualism, becomes a necessary condition of rationing. It is not in the first instance a problem of design, technique, administration, or legislation. Cultural evolution along lines already visible in developments like feminism and communalism is required.

PROPOSITION V: CORRIGIBILITY
Error is not sin.

Commentary:

a) Rapid change is typical of modern societies. Nowhere is this more visible than in the invention and discovery of treatments, methods, and techniques in health care. Rapidity of development is met by rapidity of use. Under the condition of modernity then, we must expect error to increase. We must also expect desires to be more widely broadcast and to increase in intensity. Not least of all, we must expect needs to grow more and more fluid as wants evolve into needs.

b) Under such a condition and within an historic context, distinguishing appropriate as against inappropriate moves becomes necessary. For example, it may be appropriate to expect the 21st century school to provide computers to its students but it may not be appropriate to provide parking lots, to expect every 21st century hospital to provide MRIs, etc. The question is one of timing and affordability in the context of a rationing scheme.

c) Rationing under one historic condition may be inappropriate under another and unforeseen historic condition. So, any scheme of rationing must include self-evaluation, self-criticism, and self-amendment. The conditions of transparency, participation, etc. must be fitted for corrigibility. It will not due to develop an initial rationing scheme that meets democratic conditions and then resign it to expert culture to deal with all future development [consider, in this regard, the history and present problems of Social Security and Medicare].

d) Provision for evaluation, etc. does not imply incompetence or corruption. Even under optimal conditions—which can scarcely be expected—error and failure will occur. In short, rationing calls for continuing inquiry [scientific, operational, moral] and any scheme must provide for such inquiry without being threatened.

e) In order to deal with the temptation of expert culture, rationing needs a process of appeal open to all and not limited to matters of individual use of resources. That is, at every moment and in every way possible, rationing must communicate the message of common ownership, common responsibility, and common benefit.

CODA 1: IS RATIONING POSSIBLE?

a) We have been rationing health care all along. What we have not done is admit it nor have we met criteria of democracy and justice. Thus, the poor, the unemployed, the legal and illegal immigrant, people of color, children and the young, the physically and mentally challenged, etc. have born the larger burden of scarce resources. Social stereotypes have made women "the caregiver." This unmeasured and unrewarded resource has helped sustain the system such as it is. As this "resource" vanishes as gender equality evolves, both its value is revealed and the problem of rationing is compounded. Euphemism abounds. Hardly anyone, except a maverick ex-governor like Richard Lamm, mentions the word. Denial is everywhere.

b) The very success of the modern—technology, chemistry, communication, mobility, etc.—has accelerated the need for rationing. The success of the American republic in communicating the desirability of democracy—in acting for the most part as if it really was the "city on the hill"—has generated expectations for those who never before in history expected to have expectations—at least not in this life. These expectation, moreover, are rapidly becoming worldwide. But, there is a cost attached and particularly a cost for those who have both power and wealth. On the record, we are not prepared to meet it, i.e. accept a reduction of the goods enjoyed by a global minority for the sake of a more widely shared distribution. Hence the ambiguous status of the American Republic in today's world. Consider, by way of example, the decreasing fraction of GNP that we give to foreign aid and consider too that most of what we do give is self-serving.

CODA 2: THE AMERICAN REPUBLIC— IDEALISM AND AMBIVALENCE.

a) The Republic's founders were tutored by the notion of the public good. They were shaped by the classics and idealized Rome. Today's America has deteriorated into what is called "interest group politics." But the founders were not naïve about human interest. Madison, for example, in the often-cited #10 of the Federalist Papers advocated "faction" as a way of limiting power. But, modern faction in the context of market capitalism and scientific technology has evolved into a massive inequality of factions exemplified by the political lobby as it is funded and operates in today's legislatures. As it were, the American polity has turned Madison's faction upside down.

It no longer protects the Republic; it disintegrates and even replaces it. By contrast, rationing requires a surrender of some interests in the name of the public interest. Rationing is thus inherently counter-cultural. Hence, we cannot expect faction to disappear and we can expect factions to grow fewer and stronger. Therefore, no one should be sanguine about the outlook for a just system of health care rationing in the near future.

b) We are historically and inherently suspicious of government. In that, we are all Jeffersonians, not Hamiltonians. But this too has evolved from a desirable control of power into a disdain of government per se. Sometimes genuine often only self-serving, this disdain is a useful mask for self-interest. Privatization in a market culture only pretends to political and economic virtue.

c) Apart from the evolution of faction—interest groups, lobbies, localism, states rights, etc.—American government was deliberately constructed for paralysis at worst, minimal change at best. The much praised "separation of powers," is an effective way of controlling tyranny and an ineffective way of responding to needs in a time of scarcity—no matter how needs are defined. Hence, incrementalism is ordinarily the best we can do. In health care this produces narrowly drawn and compromised legislation. For example, consider the recent attempt to meet the needs of the 10 million children without adequate health care [CHIP]. It is so hedged about with requirements and provisions and so tied to state policies that it has been of little use, at least thus far. By contrast, modern problems are systemic. Piece-meal reform not only fails but in its failure discredits reform altogether. Sneering at "do-gooders" gets its warrant not so much from their alleged sentimentality but from structural inadequacies.

d) Systemic response does occur but rarely and only in the presence of crisis. Medicare arrived in the aftermath of the Kennedy assassination, social security in the Depression, civil service out of the corruption of cities by the newly born political machines, etc. The space program was a "cold-war" response to Soviet success. Historically and ironically, most civil advances occur under the cover of military necessity.

CODA 3: IN A CASSANDRA MOOD:

I have commented on rationing as if it were only a national problem. While it is surely that, its ultimate locus is transnational and global. For example, it is possible to interpret the massive health investment of the US, and its wealth more generally, as enabled in good part by the poverty of sub-Saharan Africa, Southeast Asia, and Latin America let alone by the poverty within its own borders. Among other things, wealth creation flows from them to us as cheap labor and cheap raw materials. In other words, health care in a systemic environment impacts upon and is impacted by the totality of the environment. So, we might solve "our" rationing problem but a sense of justice might well ask how we could afford to do so. The answer, even if we should succeed, might well be very troubling!

REFERENCES

[Below are some selected texts that I have found useful in thinking about access to health care and rationing.]

Albert, T., "Medical Advances: Can Justice Keep Up?" *American Medical News*, November 25, 2002

Altman, D.E., and Levitt, L., "The Sad History Of Health Care Cost Containment As Told In One Chart," *Healthplan* 21 (2):W83 (2002)

Bellah, R.N. et al, *Habits Of The Heart,* Harper and Row, 1985

Bickenbach, J.E., "Disability, Justice, and Health-Systems Performance Assessment," in *Medicine And Social Justice*, R. Rhodes, M. P. Battin, A. Silvers, eds., Oxford University Press, 2002, pp. 390-404

Broder, J. M., "Problem of Lost Health Benefits Is Reaching Into the Middle Class," *New York Times*, November 2, 2002

Callahan, D., *Setting Limits: Medical Goals In An Aging Society*, Simon and Schuster, 1987

Congressional Budget Office, "How Many People Lack Health Insurance and for How Long?' *A CBO Paper*, May 2003

Crawford, M.J., et al, "Systematic Review Of Involving Patients In The Planning And Development Of Health Care," *British Medical Journal*, Volume 325, November 30, 2002, pp. 1263-67

Daniels, N., *Just Health Care*, Cambridge University Press, 1985

Daniels, N., "Justice, Health, And Healthcare," *American Journal Of Bioethics*, Spring 2001, Volume 1, Number 2, pp. 2—16

Daniels, N., Light, D.W., Caplan, R.L., *Benchmarks of Fairness for Health Care Reform*, Oxford University Press, 1995

Dworkin, R., *Taking Rights Seriously*, Duckworth (London), 1981

Edgar, A., "Quality of Life Indicators," in R. Chadwick, editor, *Encyclopedia of Applied Ethics* (3rd edition), Academic Press, 1998, Volume 3, pp. 759-76

Francis, L.P., "Age Rationing Under Conditions of Injustice," in *Medicine And Social Justice*, R. Rhodes, M. P. Battin, A. Silvers, eds., Oxford University Press, 2002, pp. 270-77

Galarneau, C. A., "Health Care As A Community Good: Many Dimensions, Many Communities, Many Views of Justice," *Hastings Center Report,* V. 32, No. 5, September/October 2002, pp. 33-40

Glied, Sherry, et al, "The Growing Share of Uninsured Workers Employed by Large Firms, *The Commonwealth Fund,* October 2003.

Hardwig, J., "Is There A Duty To Die?," *Hastings Center Report*, 1997, Volume 27, No. 2, pp. 34-42

Levine, C., "The Loneliness Of The Long-Term Care Giver, *New England Journal Of Medicine*, 1999, Volume 340, pp. 1587-90

Light, D. W., "The Real Ethics of Rationing," *British Medical Journal*, July 12, 1997, Volume 315, pp. 112-115

Martin, D., Abelson, J., Singer, P., "Participation In Health Care Priority-Setting Through The Eyes Of The Participants," *Journal Of Health Services Research And Policy*, October, 2002, Volume 7, Number 4, pp. 222-229

Martin,, D.K., Giacomini, M., Singer, P., "Fairness Accountability For Reasonableness, And The Views Of Priority Setting Decision-Makers," *Health Policy*, Volume 61 (2002), pp. 279-290

Maslow, A.H., *Toward A Psychology Of Being*, Van Nostrand, 1968

McGinnis, J.M., "Health In America—The Sum Of Its Parts," *Journal Of The American Medical Association*, May 22/29. 2002, Volume 287, Number 20, pp.2711–12

McGlynn, Elizabeth A. et al, "The Quality of Health Care Delivered to Adults in the United States," *New England Journal of Medicine,* 148:26, June 26, 2003

"Medicare Bill, Highlights," *New York Times*, November 25, 2003

Nelson, J.L., "Just Expectations: Family Caregivers, Practical Identities, And Social Justice In The Provision Of Health Care," in *Medicine And Social*

Justice, R. Rhodes, M. P. Battin, A. Silvers, eds., Oxford University Press, 2002, pp.278-289

Nozick, R., *Anarchy, State, And Utopia*, Oxford, 1974

Pear, R., "Panel Citing Health Care Crisis, Presses Bush To Act," *New York Times*, November 20, 2002

Rawls, J., *A Theory Of Justice*, Harvard University Press, 1971

Sen, A. K., *Development As Freedom*, Alfred A. Knopf, 1999

Sheehan, M., and Sheehan, P., "Justice And The Social Reality Of Health: The Case Of Australia," in *Medicine and Social Justice*, R. Rhodes, M. P. Battin, A. Silvers, eds., Oxford University Press, 2002, pp.169-182

Singer, P., McKie, J. Kuhse, H., and Richardson, J., "Double Jeopardy And The Use Of QALYs In Health Care Allocation," *Journal of Medical Ethics*, Volume 21 (1995), pp. 144-50

Vladeck, B. C., Fishman, E., "Unequal By Design: Health Care, Distributive Justice, And The American Political Process," in *Medicine And Social Justice*, R. Rhodes, M. P. Battin, A. Silvers, eds., Oxford University Press, 2002, pp. 102-120.

Wicks, A. C. editor, "Special Issue on Health Care And Business Ethics," *Business Ethics Quarterly*, October 2002, Volume 12, No. 4

NOTES

1. This essay is part II of a larger project. Part I, "A Trillion Dollar Problem" dealt with access to health care. It was presented at the Retreat of the Ethics Committee, South Carolina Medical Association in February 2002. *A version of this essay was published in Anthology of Recent Platform Addresses by Ethical Culture Leaders*, American Ethical Union, 2003, pp. 100-114

2. "2002 Health Care Spending Hit $1.6 Trillion," Joel B. Finkelstein, *American Medical News* (AMA), Jan 26, 2004.

3. "Deciphering the Drug Benefit," *New York Times*, December 9, 2003.

4. In fact, many of those that are uninsured do get some health care. Emergency Rooms for example are required by law to "stabilize" a patient regardless of ability to pay. Of course this care is the most expensive available because of the costs of maintaining an emergency room. Moreover, this care is usually late in an illness when things have gotten so bad that in desperation people show up at the door. Care for ordinary illness, preventive care, and the like are simply not available. And while most physicians pro-

BIOMEDICAL ETHICS

vide some free care, they cannot and do not provide it as a matter of course. Meanwhile city hospitals and community hospitals are going broke. Many are being taken over by for-profit corporations. Free clinics are closing down.

5. Parenthetic citations refer to texts listed at the end of the essay.

6. For example, consider David Riesman's "lonely crowd," Robert Bellah's "habits of the heart."

7. John F. Kennedy's inaugural, may be read as a last desperate call for community or else as a cruel commentary on its absence. I wonder how "ask not what your country can do for you but what you can do for your country . . ." echoes in our minds today. Perhaps the nostalgia for "Camelot" reflects as well the nostalgia for the lost community.

13.

Poverty, Health, and Bioethics

Robert B. Tapp

In framing my theme, I will need to move back and forth between the values of the Enlightenment and those of our own time. In dealing with the values that shape my approach to that theme in the contemporary situation, I will suggest a major reorientation of values focusing not at the "top end" of the scale where we usually work but rather at the very bottom.

John Locke, as almost everyone knows, argued that democracy was founded upon the values of "life, liberty, and property." Jefferson replaced that third value, "property," with "the pursuit of happiness." A focus on happiness, of course, echoes of Aristotle's *Nicomachean Ethics* so Jefferson's was an Aristotelian expansion, as it were. Certainly, the change has resonated more favorably among Americans than would have "property" in the time since 1776.

The meanings of this broad value framework have been expanded in the ensuing years. To be sure, it required a devastating civil war, but "liberty" eventually was interpreted most dramatically to apply to slaves and not just to "free" white men. The process, as in the civil rights movement expands on the meanings and uses of liberty and, of course, still continues. The value most germane to my topic, however, is "life." In the eighteenth century this referred to the necessity of a trial before capital punishments could be carried out. From this relatively narrow base, "life," as we know, has become a far more complex and inclusive moral, political, and legal term.

Words such as life, liberty, happiness function as signals of values—to be re-defined, expanded, revolutionized. Our political debates should remind us, however, that social life requires us to recognize that values often compete. Ideologically and politically, we often find ourselves divided on ways to balance such competing values.

It would be helpful at this point in dealing with value competition to recall the work of the great social psychologist Milton Rokeach. Beginning his career with researches into attitudes (their distributions, shifts, causal underpinnings), he realized that human beings create for themselves hierarchies of conceptual organization that help order their experience. Eventually he designated the levels of a cognitive pyramid as "Values, Beliefs, and Attitudes." A small set of values, he argued, subsumed a larger set of beliefs that in turn subsumed yet larger groups of attitudes. Why not focus then on the high-level ordering-set of values in our attempt to understand different human behaviors?[1]

This question led to a detailed study of instruments that social psychologists had used to probe and classify values. Rokeach finally concluded that values could be grouped under headings of "terminal" (what a person desires—"end-states") and "instrumental" (ways to achieve these desires—"means").[2] It is important to note that these were all "positive" values. Eventually Rokeach extracted 18 terminal and 18 instrumental values from his empirical studies. Subjects were then asked to rearrange each in terms of their own preferences. The mean rankings of different groups of subjects could then be compared. Much of the reported research by Rokeach before his untimely death was with the two terminal values of "freedom" and "equality."

Rokeach had been bothered by the researches on the "authoritarian personality" done by Max Horkheimer and Theodor Adorno during and after World War II. These émigrés from Nazi Germany had understandably been concerned with determining the psychological markers of fascist personalities. Rokeach, as a US social democrat, was equally determined to delineate the values structures of Stalinists and their sympathizers. His findings led to a fourfold typology based on the relative rankings of freedom and equality. To sum up, for capitalists, freedom was high and equality low. Among communists, freedom was low and equality high. For fascists, both were low; and for socialists, both were high.

Labels and rhetoric become very important here. For example, I cannot imagine anything like "socialism" ever becoming desirable to a majority of the US citizenry. Stigmatized by Nazi and Soviet misappropriations of the term, and smothered by the obfuscations of right-wing money, the think tanks it supports and the politicians it helps elect, that term is not realistically available in America. In a similar vein, some have recently suggested that "liberalism" is also so tainted a term that we should revive "progressive." But then could we ever forget the rhetorical excesses of the Progressive Citizens of America and the Henry A. Wallace campaign in 1948?

My own suggestion: we should revive talk about the "common good" and the "commonwealth." These terms would be much harder for the right wing to target. They have longstanding historic and religious roots and can easily embrace the democratic and Enlightenment values of humanists and many of their fellow citizens.

Given these and other concerns about the effects of language and usage on values, our times call out for more studies like Rokeach's of the effects of religious as well as secular ideologies on values. Of importance here is an even more lasting insight of Rokeach's. We must study the relative meanings and rankings of various values in order to understand what is going on. We will then realize that a good portion of the meaning of a value lies in how it is related to other values. This certainly holds for the three Enlightenment values of life, liberty, and the pursuit of happiness.

Let me focus now on that first value, "life." What has it come to include in our times—and how can these newer, broadened meanings be implemented? In replying, I know that I run the risk of being confused with the sectarian uses of "right to life." But, I do not want to surrender the idea. I want to treat the "right to life" as a "human right" that is based on the rights of humans to exist—and to exist over against controllable violences that threaten or destroy that existence.

UNIVERSALIZING THE
BROADENED MEANINGS OF "LIFE"

If in our time we interpret "life" to include access to life-saving medicines, this is a fairly recent vision. As James Bryant Conant once
argued, until about 1900 medicines killed about as many people as they
cured.[3] Today, however, the situation is entirely different. We indeed
have medicines that cure and/or prevent many life-threatening diseases.
We also have a growing body of public health knowledge that makes it
possible to prevent or reduce many of the ills that have shortened
human lives in times past, i.e.cleaning up water supplies, disposing of
sewage and toxic waste, cleaning up air pollutants, etc.. I call them
"knowledges" rather than "experiences" since they illustrate the ways
that the sciences are embedded in political and economic realities.

In the middle of the past century, at the close of the bloody World
War II, a *Universal Declaration of Human Rights* was created under the
leadership of Eleanor Roosevelt. I would focus on two of its articles,
numbers 25 and 27, and in particular on the phrases that I have put in
boldface. The Declaration was adopted in 1948, but, as we know,
nations are still very far from implementing many of its provisions.

Article 25

1. Everyone has the right to a standard of living **adequate for the
health and well-being of himself and of his family**, including food,
clothing, housing and **medical care** and necessary social services,
and the right to security in the event of unemployment, sickness, disability, widowhood, old age or other lack of livelihood in circumstances beyond his control.

2. Motherhood and childhood are entitled to special care and assistance. All children, whether born in or out of wedlock, shall enjoy
the same social protection.

Article 27

Everyone has the right freely to participate in the cultural life of the
community, to enjoy the arts and **to share in scientific advancement and its benefits**.

OUR PREOCCUPATIONS WITH VIOLENCE

The phrasings in these Articles, of course, had many precedents. My point here is that 55 years later we have made only the most limited progress in realizing them either locally or globally. No doubt, there are extenuating circumstances as events have intervened to make implementation difficult. The Cold War diverted much research energy and resource into improved deliveries of violence on a global scale. Countries were drawn into this polarized struggle, and poorer countries were forced to divert even larger proportions of their gross domestic product (GDP) into violence and its means. I add that this term covers all those other terms that mask their violent nature, i.e. that are called "defense, " counterinsurgency," "state security," "regime change," "covert actions," and the like. I deliberately avoid the term "terrorism," not because it is innocent of violence, but because it is ambiguous and inevitably relativistic. More often than not, what one group calls "terrorism," another other calls "liberation struggle." Violence, however named, limits, or threatens to limit, lives, liberties, and properties. Any moral evaluations of it in its many forms must start with that empirical fact. Even the exercise of police powers within a society can be understood under certain conditions as a form of violence.

Suppose, however, that we take the Declaration seriously and argue that the "right to life/right to exist" is the most basic of all human rights. We will then be in a position to examine those forces that negate this right. The broadest term for such forces is violence. First, we will need to distinguish between personal violence and structural violence. When A decides to kill B, or destroys his or her liberty or property, that is clearly violence and most cultures recognize it as such. Suppose that the community decides that someone who steals from another should be tried and incarcerated. Liberty has been lost, and there is a sense in which we must say that violence has therefore occurred. More accurately, one act of violence has generated another. When such acts are built into law, they acknowledge personal violence and claim the need for a violent response to it. Faubion suggests that the application of moral judgment to violence in this most general sense is best explained by using the relatively neutral terms of "moral agent" and "moral patient."[4] Thus the agent is the initiator of violence; the patient is the recipient of it. Their relationship is characterized as violent.

STRUCTURAL VIOLENCES

In addition to these intended situations of violence, we need to assess the roles of structural violences. The root of the concept is in a neo-Marxian insistence on the analysis of collective forces. In our time, liberation theologians (predominately but not exclusively Roman Catholic) have been most active in developing such analyses. The Encyclicals of John XXIII as well as the Medellin conference of Latin American bishops (1968) centered on the concept of a "preferential option for the poor." The critique of liberalism (both the North American versions and the more worldwide "liberal" economics), building upon a classic conception of "common good," emphasized increasing inequalities as the inevitable outcome of individualist ideologies.

POVERTY AS THE ROOT

If being poor—existing in poverty—is an inevitable part of the economic system, then all the consequences of poverty are predictable; none are "accidents." They are not the results of ignorance or laziness. Increasing economic polarizations are "internal to the system and a natural product of it."

What becomes even more interesting in our times is that the situation of the former Third World has now been globalized. Poverty may be somewhat relative to time and place (the economists have spoken at length about "relative deprivations") but increasing gaps between those at the top and those at the bottom are easily quantifiable and have predictable political consequences in a world of widespread communications.

These considerations clearly shed light on a "right to life/right to exist. In turn, this requires us to grasp the relationship poverty to biomedical ethics. In its simplest form: who gets treated and for what? Of course, there are other variables that enter the equation, i.e. gender, age, ethnicity, social class, religion, citizenship, sexual preferences. These all are affected by and affect social power and therefore shape decisions in biomedical research attention and resource allocation.

The benefits of medical science are allocated on the basis of social power. Most of us in the "developed" world see the distributions of power as relatively just. The misfortunes of individuals are seen by liberals nearly as much as by conservatives as accidents and not as parts of structures that need correction.[5] As Robert McAfee Brown, a Protestant liberation theologian, paraphrasing Jon Sobrino, put it, "unless we agree that the world should not be the way it is . . . there is no point of contact, because the world that is satisfying to us is the same world that is utterly devastating to them."[6]

Paul Farmer, a physician-specialist in infectious diseases and an anthropologist, observes that "it is unnerving to find that the discoveries of Salk, Sabin, and even Pasteur remain irrelevant to much of humanity."[7] He has emphasized that we should approach human suffering on a global scale, and apply scientific-medical knowledge globally. He reminds those of us dedicated to Enlightenment values that Adam Smith approached ethical situations from the position of "impartial observer." John Rawls' "veil of ignorance" is the contemporary version of such a stance.

The implications of the *Universal Declaration on Human Rights* for health care had received rather casual treatment since 1948. Consequently, a group of care-providers gathered at Tavistock (London) in 1999 and came up with five principles to expand upon health as a human right. These were grew into seven principles at a 2001 conference at the American Academy of Arts and Sciences:

Rights: People have a right to health and health care

Balance: Care of individual patients is central, but the health of populations is also our concern

Comprehensiveness: In addition to treating illness, we have an obligation to ease suffering, minimize disability, prevent disease, and promote health

Cooperation: Health care succeeds only if we cooperate with those we serve, each other, and those in other sectors

Improvement: Improving health care is a serious and continuing responsibility

Safety: Do no harm

Openness: Being open, honest, and trustworthy is vital in health care[8]

All three, the Declaration, Tavistock, the Academy put it very clearly. The right to health is a basic human right—and nations have failed miserably to recognize this right. The fact that Libya and Syria are members of the UN's commission on human rights illustrates how problematic this has become. The risk, of course, is that "human rights" simply becomes a mantra, to be intoned but trivialized.

HEALTH AS THE BASIS OF THE RIGHT TO LIFE

Those of us who are committed to organizations of moral articulation and moral education need to start here. The right to be a person of any color, and/or to be a woman, and/or to be gay-lesbian-bisexual-trans-gendered is conditional upon health. And health must be understood as a social-communal matter. HIV-AIDS, STDs may command attention because of presumed relationships to sexuality—but TB, SARS, and many viral plagues move freely in our air, water, and sewage systems—infecting millions and affecting all of us.

RECOGNIZING AND FINANCING HEALTH

Prevention of illness is in most cases easier and much less expensive than cure. So public health measures, adequate diet and housing and the like are instrumental to the "right to live/right to exist." Since many of the causes of disease are already understood and treatments available, untreated persons are deprived of their human rights. Thus, it is not acceptable that, in many countries with "universal" provisions for some kind of health care, non-citizens are excluded. Further, today's wars in the Middle East and sub-Saharan Africa have created large populations of refugees who are usually left out of whatever health systems are enjoyed by their hosts.

Even in the wealthy US, more than 40 million persons at any one time are without health insurance and so without dependable health care. Most of those who have insurance depend upon steady employment, a diminishing condition in the present economic picture. When the costs of pharmaceuticals are included, the insurance coverage becomes even more tenuous. The large well-organized lobbies of highly-profitable pharmaceutical companies killed the attempt at universal health care during the beginnings of the Clinton administration. Those forces are surely stronger today than then. Given the rightward shifts of US voters, it seems unlikely that even patchwork fixes will actually extend any "right to health and health care."

THE MORAL PRIMACY OF HEALTH

If the present systems are unfixable, the best place to start developing alternatives would be in a country with ample resources of funds, providers, and intellectual/moral talents. This probably means the United States. Strategically, we will need to assemble a massive lobby, on a world basis, to push for a recognition of the basic and universal right to health care as a requirement of the right of all humans to live. For this to happen, the resistance and rejection already demonstrated many times and predictable must be challenged.

The record is not good. As I write, US policies remain ambiguous in terms of allowing poorer nations to import generic forms of drugs. Thus clearly-treatable diseases such as malaria and TB remain superimposed on a poverty map. The current US administration has not only rejected the Kyoto accords on pollution and global warming but advanced regulations allow polluting older factories to expand while evading environmental cleanup.

The urgency of redirecting these policies, based on the analysis of present failures, comes from a distinguished neuropharmacologist, Floyd E. Bloom:

So when I gave my speech as the outgoing president of the A.A.A.S., I decided to use the opportunity to call for a national commission to restore the American health system. The idea is to get patients, providers, insurers, employers, caregivers and physicians together to think about the future of medicine.

Frankly, I'd like for us to consider health care to be regarded as something like a public utility. To me, if we agree that universal coverage is something to be desired, is that really much different than the fact that we've all agreed that everyone in the country is entitled to have electricity, water, telephone connections, if they can pay for it. We have all kinds of ways to help people get those basic provisions of life.

And health benefits could be viewed in exactly that same utilitarian way. It could be a corporate network like water power and electricity, with regulatory agencies that set the rates for profit.[9]

As a right to life/right to exist identifies a right to health care as basic to its realization, others instrumental rights will be called to our attention. For example, better health care will, by lowering maternal and infant mortality, increase population pressures. In turn, expanding and improving educational opportunities will be needed to deal with these. The declining quality of our educational systems needs to be reversed. An increase of effective education could modify the authoritarian trends visible in our country. Additionally, with education will come more researchers in all the sciences. Rights, in other words, reflect a tissue of social connection, again, the Commonwealth.

PHILOSOPHICAL CODA

In the 1960s, when I was trying to function as a Unitarian Universalist (UU) theologian, I created the phrase "the expansion of the quality of life" to articulate something beyond simply the presence or absence of physical existence. Such a focus seemed appropriate when addressing the affluent and highly educated liberal religionists I was dealing with. For a variety of reasons, not of interest here, that phrase never succeeded in the Universalist Unitarian circles. But it did, however, begin to appear within a few years in American political discourse—even from the White House.

In more recent years "quality of life" has surfaced in economic circles, complementing if not displacing a quantitative focus on monetary policy.[11] The gross product of an economy, and equally easy to display a comparison of averaged GDPs. But these numbers can be quite deceptive. Imagine the average income of the board of directors of the Microsoft Corporation before and after Bill Gates enters the room.

I argued then that Unitarian Universalists (and Ethical Culturists) would do best if they explored and developed their own moral/ethical discoveries. They were (and still are) highly educated and wealthy persons living in advanced societies without the confounding disabilities of theisms, moralities, and ideologies handed down from agricultural pasts.

The rise of fundamentalisms around the world, and their enormous success in the United States, makes those goals still relevant. For some time, economists have reminded us of the distinctions between income and wealth. In the present situation of high unemployment, that distinction becomes much more poignant. But the increasing inequality of income and of wealth within this economy, as well as within much of the developing world, makes my humanist agenda something of a "luxury" (however essential it may be in the long run). Our "right" to develop and promulgate the lessons of our experiments in post-traditional living must be joined with our willingness to simultaneously develop and promote an agenda for the rest of humanity that will effectively reduce poverty and its devastating results.

There is indeed a hierarchy of human needs that cannot be ignored. Life, health, shelter, food come first! Freedoms of thought, speech, actions can only be sustained on that material base.

NOTES

1. Rokeach, Milton. 1968. *Beliefs, attitudes, and values; a theory of organization and change.* San Francisco: Jossey-Bass.

2. John Dewey had termed these Intrinsic and Instrumental in 1916 in *Democracy and Education.* He had also argues that value pointed not only to something that was prized and cherished but to the "evaluating" process of comparing and ranking values. He argues that democracy would have "its

consummation when free social inquiry is indissolubly wedded to the art of full and moving communication." (*The Public and its Problems* [1927] in *Later Works* 2.350).

3. Conant, James Bryant. 1952. *Modern science and modern man.* New York: Columbia University Press.

4. Faubion, James D. 2003. "Religion, Violence and the Vitalistic Economy." *Anthropological Quarterly* 76:71.

5. Pixley, Jorge V and Clodovis Boff. 1989. *The Bible, the church, and the poor.* Maryknoll, N.Y.: Orbis Books.

6. Farmer, Paul. 2003. *Pathologies of power: health, human rights, and the new war on the poor.* Berkeley : University of California Press. A general appreciation of Farmer's vision can be found in Kidder, Tracy. 2003. *Mountains beyond mountains.* New York: Random House.

7. Farmer, op. cit, p. 157.

8. Ibid, p. 144.

9. Ibid., p. 319, n. 16.

10. Dreifus, Claudia. 1903. "A zealous quest for chemicals to heal ailing brains." in *New York Times.*

11. Nussbaum, Martha Craven, Amartya Kumar Sen, and World Institute for Development Economics Research. 1993. *The Quality of life.* Oxford England: New York: Clarendon Press; Oxford University Press.

Addendum

14.

Before and After Schiavo
Thirty Years of Bioethics, Autonomy and Surrogacy

Carmela Epright

O n March 31, 2005 Terri Schiavo died of multi-system failure, nearly two weeks after her feeding tube was removed, and more than 15 years after her collapse from a cardiac arrest brought about by an eating disorder induced potassium imbalance. The Schiavo case sparked unprecedented public debate and was responsible for a dizzying array of court cases, lawsuits, rulings, and stays (including three Supreme Court and two gubernatorial rulings, and act of congress). The case continues to spark both public and scholarly debate concerning the proper appointment of surrogate decision-makers, the withdrawal of care, and the status of nutrition and hydration as medical treatment. In what follows I will discuss the Schiavo case in the context of the history of bioethics by outlining the three decades of legal and moral debate that preceded this case and argue that, given this context, the Schiavo case was (ultimately) rightly decided: the proper surrogate made the decisions regarding Terri's care, and the withdrawal of Mrs. Schiavo's feeding tube was appropriate—indeed it was long overdue.

In order to understand the legal decisions made in the Schiavo case it is essential to examine a few of the "watershed" bioethics cases that formed the foundation for the legal and moral decisions made in the case of Terri Schiavo. The first of these cases, that of Dax Cowart, set the stage for discussion of patient autonomy—the primary principle that underpinned the need for a surrogate decision-maker in cases like Schiavo's. Next, I will discuss the cases of two women declared—like Terri Schiavo—to be in a type of permanent unconsciousness called a persistent vegetative state. The Karen Ann Quinlan and Nancy Cruzan cases hinge specifically on the role of surrogate decision makers in situations in which a patient is unable to articulate his or her desired level of medical intervention.

DAX COWART[1]

In the summer of 1973 Donald "Dax" Cowart was critically injured in an explosion that killed his father. Dax sustained second and third-degree burns over more than sixty-five percent of his body. From the very beginning he clearly expressed a desire not to be treated for his injuries. In every instance where his consent was sought he refused to grant it. Eventually his health care providers stopped asking—they assumed that the pain associated with his treatment made him incompetent to make decisions about his care, so they ignored his demands that treatment be terminated. At the time the only treatments available to persons with burn injuries were excruciatingly painful—they required daily "debreting" sessions, whereby the patient is submersed in a chemical bath while health care providers scrape dead and infected tissue from their body with sharp instruments. Because the accident rendered him significantly disabled—completely blind, partially deaf, fingerless and unable to walk without assistance—he could not leave the hospital on his own accord. Thus he was forced to undergo whatever treatments his health care providers deemed necessary—regardless of how painful or invasive they might be.

During the time that he was hospitalized multiple psychiatrists concurred that he was competent according to the standard criteria. Thus Dax argued that his treatment should have been terminated when—as a competent adult—he asked to leave the hospital and return home to die from his injuries. This is a position that he steadfastly holds to this day, despite the fact that he went on to marry and become a successful attorney.

Although "Dax Cowart" is not a household name, like Karen Ann Qunilan, Nancy Cruzan, and Terri Schiavo—his represents the first of the watershed cases in what became known as "The Right to Die" movement. Today the doctors and other health care providers that treated Cowart against his will would not only be civilly liable—they could be charged with aggravated assault, criminal battery, and perhaps even kidnapping. As I outline below, there are many lingering questions about the rights of individuals in a health care setting, but it would be a mistake to ignore the tremendous progress that has been made over the last three decades. Today a competent adult[2] cannot be held—by either physical force or by the coercive employment of his or her illness or disability—against his or her will. He or she can *at any time*, refuse or terminate *any* medical treatment, even if that treatment is necessary to preserve his or her life. Moreover, a competent adult's self-regarding health care decisions cannot be overridden—by his or her spouse, children, pastor or doctor—even when such people are motivated by what they take to be the patient's "best interest."

The assumption that adult patients are rational beings (unless proven otherwise) and thus that their self determination—their autonomy—ought to be respected is one of the great contributions of bioethics to medicine. As one can see from the Cowart case, it was not long ago that medicine operated under the opposite assumption that because patients are sick and do not have as much information about their health as do their doctors, they are not rational enough to make decisions on their own behalf regarding their care. The overriding ethos was that patients should be treated with paternalism—the benevolent protection provided by their wiser, more experienced doctors. In the last 30 years, since cases like Dax Cowart's came to the attention of the general public, the pendulum has swung in the opposite direction—patients are seen to be the absolute arbitrators of their own interests in a health care setting.[3]

Understanding the primacy of patient self-determination is absolutely essential to comprehending the way in which the Schiavo case unfolded. That this is true is best explained by examining the cases of two young women who, like Teri Schiavo, existed in a persistent vegetative state, entirely unaware of their place at the center of controversy and the nationwide attention.

KAREN ANN QUINLAN

Although it continues to be widely reported that Karen Ann Quinlan's cardiac arrest on April 15, 1975 was brought about by combined use of drugs and alcohol, this claim was not confirmed by a toxicology exam. What is known is that after she collapsed at a party, friends and fellow partygoers unsuccessfully attempted to resuscitate her. She became unconscious; and, over the course of the next several months fell into a persistent vegetative state—a form of permanent unconsciousness characterized by irreversible cessation of cognitive functions of the brain while brain stem functions, such as breathing, remain intact. The controversy surrounding Karen Ann Quinlan revolved around questions of surrogate decision-making.

In Karen Ann's case her appointed guardians, her parents, Joseph and Julia Quinlan wanted to remove Karen from the respirator that doctors insisted was necessary to keep her alive. The Quinlan's argued that they knew Karen Ann's wishes better than any member of the hospital staff—all of whom were strangers to this patient prior to her collapse on April 15th and none of whom had known Karen Ann in a conscious state. Her parents testified that their daughter would not wish to be artificially sustained using what they deemed to be extraordinary measures.

For their part hospital officials petitioned the court for guardianship of Karen Ann, arguing that the care that they wished to provide was consistent with this patient's best interests. Although the Quinlan's lost the first case in District Court, the New Jersey Supreme Court issued a precedent setting ruling declaring that when the wishes of an adult regarding his or her care can be established, that individual's wishes must be upheld. However, how one goes about establishing those wishes—even with an advance directive (which Karen Ann did not have) remains unclear. Even people who are familiar with the workings of modern medicine cannot fully predict all of the procedures that might be necessary for their care and family members may well subscribe to different values than the patient. Questions of this sort were re-introduced to public debate through another tragic case involving a young woman.

NANCY CRUZAN[4]

In January 1983, twenty-five-year old Nancy Cruzan lost control of her car as she traveled down a Missouri road. The car overturned, and she was discovered, lying face down in a ditch, without detectable respiratory or cardiac function. It was estimated that she was oxygen deprived for twelve to fourteen minutes. (Permanent brain damage generally results after four to six minutes without oxygen.) Emergency medical technicians were able to restore her breathing and heartbeat, and she was transported to a hospital in an unconscious state. Cruzan remained in a coma for approximately three weeks, and then progressed to an unconscious state in which she was able to orally ingest some nutrition. In order to ease feeding and further her recovery, surgeons implanted a feeding and hydration tube In October 1983, at which time she was admitted to a state hospital. However, subsequent rehabilitative efforts proved unsuccessful. It became apparent that she had virtually no chance of regaining her mental faculties, and her parents—who had been appointed as her co-guardians—asked the hospital to terminate the medically assisted nutrition and hydration procedures. The hospital refused to honor the request without court approval.

The court decisions revolving around the Cruzan's battle with the State of Missouri are long and complex. The state argued that that (1) Although Cruzan was in a "persistent vegetative state," she was neither dead within the meaning of Missouri's statutory definition of death nor terminally ill thus her treatment was of "compelling interest" to the state; (2) Cruzan's right to refuse treatment did not outweigh Missouri's strong policy favoring the preservation of life; (3) Conversations with her friends and family members concerning her desire to not "be placed on machines," unless she "could live at least halfway normally" were unreliable for the purpose of determining her intent, and were thus insufficient to support the parents' claim to exercise substituted judgment on her behalf; and, (4) No person could assume the choice of terminating medical treatment for an incompetent person in the absence of "clear and convincing, inherently reliable evidence," which they originally claimed were absent in Nancy Cruzan's case. The United States Supreme Court agreed to hear the Cruzan case in December 1989. For the most part this court upheld the state of Missouri's claims. It found that due process was not violated by the Missouri requirement that an incompetent person's wishes with regard to the withdrawal of life-sustaining treatment be *proved by clear and convincing evidence*, and the court held that such evidence is a standard to be set by states. In the end, however, further witnesses satisfied Missouri courts that such clear and convincing evidence of her wishes *did* exist, and medically assisted nutrition and hydration was terminated.

Although the ruling in *Cruzan vs. Missouri* seemed to be a set back for the "Right to Die Movement," it resulted in a number of important precedents that ultimately benefited the movement. It also raised new questions. First and most importantly, the Supreme Court determined that the standards for surrogacy are to be established by the state. Following the ruling in Cruzan nearly every state passed statutes that outlined the ways in which surrogates would be appointed and how decisions regarding patient care would be made. Following the Quinlan precedent most states determined that the wishes of the patient—when they can be established—were to serve as the criteria for surrogate decisions. Furthermore, most states determined that surrogate decision-making ought to be performed by the people most likely to know the patient. Importantly, with reference to the Schiavo case, nearly every state (including Florida, where the Schiavo's resided) determined that when a spouse existed this person would become the surrogate, whereas the parents of patients are to be consulted if and only if the patient has neither a spouse nor adult children.

Another significant issue that emanated from the Cruzan case had to do with the medical status of delivering artificial nutrition and hydration. The life of a patient in a Persistent Vegetative State can be sustained for years through nutrition and hydration delivered by intravenous or surgically implanted tubing. Clearly, medical expertise is necessary for the delivery of this kind of nutrition and hydration. For this reason courts, have treated medically assisted nutrition and hydration as a form of health care. Thus, like all other forms of health care, it can be refused and withdrawn if its delivery is refused by or against the will of the patient.

How do we determine what Terri Schiavo would have wanted if she were making decisions for herself? We can't. It is likely that as a young, healthy woman she never seriously considered what she would want if she were to become persistently vegetative.

But we do know one thing with certainty—she chose her husband. At the time of her cardiac arrest she had been married to Michael Schiavo for four years. By marrying him she endowed him with decision-making capacity over many aspects of her life: in nearly every jurisdiction a spouse is the sole arbitrator of all financial and familial decisions when the other spouse is incapacitated (unless another decision maker has been appointed). The Supreme Court established in *Cruzan* that states could determine the ways in which surrogates are chosen and the standards that those surrogates must meet. The state of Florida established that unless one identifies (through an advance directive) a different surrogate decision-maker, the legal spouse becomes the surrogate. Thus the courts had no choice but to assume that Terri's marriage to Michael constituted an endorsement of his judgment as her surrogate.

It might very well be possible to argue that the Schiavo marriage was not a happy one, but even if this were true, it is not the state's job to adjudicate family life. People have many different kinds of relationships, and in the absence of legally provable abuse or fraud, states do not—and ought not—question a person's choice of marital partner or marital situation. It should also be noted that in the course of 7 full trials and 26 hearings—fraud and abuse on the part of Terri's husband Michael was never proven. Indeed Terri had been in a persistent vegetative state for 12 years and her case had been heard in open court multiple times before Terri's parents accused Michael Schiavo of abuse and neglect of Terri during their marriage. These charges were not proven. Thus Michael Schiavo was recognized to be Terri Schiavo's guardian and surrogate. This decision was frequently challenged, yet each court upheld his status in every hearing and trial.

Terri Schiavo's case was tragic—as are all such cases—but it was unusual only insofar as it resulted in such familial acrimony and public attention. Cases like Mrs. Schiavo's are quietly agonized over, discussed and decided everyday—in the 13 years I have worked as a hospital ethicist I have mediated more than a dozen similar decisions.

Perhaps the most significant lesson to take away from this case is the recognition that the principles that those of us in health care community rely upon to adjudicate such cases: autonomy, self-determination, informed consent, the right to treatment, the right to terminate treatment, and the right to have one's wishes interpreted and articulated by a surrogate, are neither well known nor well understood in the larger world. This case should teach us that we bioethicists have to stop talking to ourselves and do a better job initiating these conversations with the general public. I think that we clinical bioethicists do a good job at the bedside—I for one feel proud of the work that I do with families in crisis; but I would like to see us help more families *avoid* such crises—to think about and talk about these issues with one another before they themselves or their loved ones are embroiled in such controversies.

NOTES

1. There are scores of articles, books and commentaries written about the Cowart case, I would recommend starting with *Dax's Case: Essays in Medical Ethics and Human Meaning*, edited by Lonnie D. Kliever, Southern Methodist University Press, 1989. One might also want to view the video "Dax's Case" (New York Filmakers Library, 1984), which includes actual footage of the treatments endured by Cowart.

2. A patient's decisions can be overridden if he or she is declared legally incompetent or incapacitated—standards that vary widely from state to state.

3. Personally, I would hold that there are ways in which this single-minded emphasis on patient autonomy has become problematic, I address this view at length in my paper entitled, "Bioethics and Justice: Economics, Care and Conflict" published in this volume.

4. For an excellent, informative discussion of the Cruzan case, see Bill Colby (Counsel for the Cruzan family during the various trials surrounding the case), *The Long Goodbye*, Carlsbad, CA: Hay House, Inc., 2002.

15.

Prescription Drugs, Medicare Part D

Kristy Maher

T he Medicare Prescription Drug, Improvement and Moderniza-
tion Act, or Part D of Medicare, was signed into law in 2003.
Hailed as the biggest overhaul of Medicare since its inception, the program that began January 1, 2006, has faced much criticism. While
some state that the "D" stands for "drugs," critics say it should stand
for "disaster."[1] Several concerns arise from the suspicious way in
which it was passed to the problems with enrollment and its gaps in
coverage.

In some ways, this plan seemed troubled from the beginning.
Many votes cast in favor of this plan in the Congress were made with
false information regarding the cost of the program. Specifically, its
cost is far greater than originally projected. When initially signed into
law in 2003 the estimated costs were $395 billion. Within a couple of
months after it was passed, the White House recalculated and raised
the estimate to $534 billion.[2] These changing figures were not simply
the result of "miscalculations." The estimate was updated again in
February 2005 to $720 billion.[3] According to Oliver et al.,

In March 2004 the chief actuary of the Centers for Medicare and
Medicaid Services revealed that as early as the previous summer, his

office had estimated much higher costs for the proposed reforms than congressional budget analysts had. His superiors in the Bush administration, however, ordered him to withhold the estimates from members of Congress and warned him that "the consequences for insubordination [would be] extremely severe."[4]

In short, Medicare Part D passed in the House in the dead of night by one vote. The Bush administration promised doubting Republicans that it would cost no more than $400 billion over ten years. The estimated cost is now almost twice that.

A second area of concern is the complexity of the enrollment process. While choice is good, too much choice can be overwhelming. Despite the fact that assistance is available both on line at www.medicare.gov and in written form, with the 92-page booklet entitled "Medicare and You," for many the research required to successfully select a plan is daunting. Individuals must first figure out where they fit in the grand scheme of things and what options are available to them. Candidates for the program may come from a variety of existing plans including current recipients of the original Medicare plan only, persons with Medicare and Medigap supplemental insurance, retirees with employer/union-provided prescription drug coverage, those enrolled in the Medicare Advantage Plan and individuals currently receiving both Medicare and Medicaid.

Once a person's current standing is determined he/she faces a plethora of choices that vary substantially from one area of the country to another. Choices must also be made from a variety of options taking into account cost, coverage and convenience. Not all pharmacies accept all plans. Not all plans cover all drugs. Each plan has a "formulary" or list of drugs and there is no guarantee that the drugs one needs are on the list. Additionally, these lists can change every year, so the drug coverage and "co-pay" that one has upon enrollment may change from year to year. As plans evolve and change, reevaluation of coverage may need to occur on a yearly basis to assess whether or not one's current plan is still the best choice. Like Medicare Part B, Medicare Part D comes with built-in penalties for signing up after the initial enrollment period ending May 15, 2006 although this policy is now [May 06] being challenged in the Congress. In short, according to some critics, the biggest problem with the legislation is its complexity.[5] To select the best plan people need to know how often they plan to get sick, what drugs they will need to take, how much of those drugs they will need, how much insurance companies pay for those drugs and how much they can afford to spend. Many of these details obviously cannot be known in advance.

Even after a plan is selected, however, there are significant gaps in coverage both financially and in terms of which drugs are covered. Certain drugs are simply not covered by any of the available options. Additionally, regardless of the plan one chooses, there are expenses involved. Premiums vary widely based on coverage and most plans charge a deductible for prescriptions. Under the standard Part D program, enrollees pay a $250 deductible, and then 25% of prescription costs up to a specified limit.. As such, while the plan pays seventy-five percent of the cost up to $2,250, it pays nothing above that amount until drug expenses go up to $5,100).[6] This gap is sometimes referred to as the "doughnut hole" since you get something on each side, but nothing in the middle.

Once a person spends $5,100, he/she then pays 5 percent for all additional drug costs for the remainder of the year. This may result in some beneficiaries spending more than they spent before the so-called drug "benefit" went into effect. If you require a large number of prescriptions, or a small number of expensive prescriptions, the costs could be devastating.[7] According to Quadagno, the "doughnut hole" was a concession to the American Association of Health Plans (AAHP), which represents managed care firms.[8] Since, in the past, many of the elderly who chose Medicare HMOs over the traditional Medicare program did so mainly for their prescription drug coverage, managed care organizations stood to lose customers to Medicare Part D. If Medicare Part D assumed all drug costs, HMOs would be a less attractive option. Consequently, negotiations with AAHP resulted in a benefit that paid some of the costs for low spenders, covered most of the costs for people with catastrophic drug expenditures and preserved the free market for the middle class (Weissert 2004).[9] The "doughnut hole" results in the need for some people to continue to purchase additional coverage.

The larger issue affecting this plan is the high cost of pharmaceuticals. Prescription drug spending has been the fastest growing component of the health care bill over the last couple of decades.[10] While this is related to a higher prevalence of chronic illnesses treatable with pharmaceuticals, there are more sinister explanations for the costs. Many argue that the government's unwillingness to negotiate the cost of prescription drugs has only exacerbated the situation. The 2003 drug bill specifically *prohibits* the government from engaging in negotiations over drug prices. This is the same hands-off concession that was granted to providers in exchange for their cooperation in Medicare.[11] This feature resulted from the efforts of the 620 lobbyists of the Pharmaceutical Research and Manufacturers of America (PhRMA). In the first six months of 2003, the PhRMA pumped $8 million into a lobbying campaign against price controls.

This arrangement is unique. All other industrialized governments negotiate over drug prices with drug manufacturers. Additionally all major health care plans in this country have negotiated over drug costs for years. In fact, negotiations with pharmaceutical companies are not unprecedented for the U.S. government either. It has used such tactics for the Veteran's Administration (VA) system for years. Examination of the VA system highlights the savings that can be gained by negotiating over drug costs. In fact, preventing Medicare from negotiating with pharmaceutical companies seems to fly in the face of a basic purchasing principle: Buy in large quantities and get a lower price. The government's prohibition on negotiating drug costs for Medicare-Part D raises suspicions about inappropriate relations with big business. It seems the real winners from the Medicare Part D are the pharmaceutical companies who stand to reap windfall profits from inflated drug costs.

In sum, Medicare Part D, an attempt to bring prescription drug coverage to millions of Americas, while seemingly put forth for the benefit of older Americans, seems to be benefiting some far more than others. Not only was it passed under a false assumption about its cost, it was designed to benefit pharmaceutical companies with the no-price-negotiation feature and managed care organizations with the "doughnut hole" feature. Its complexity leaves even the most resourceful individuals mired in a plethora of plans where the "best" decision is not obvious and reevaluation of which plan is "best" may need to be made on a yearly basis.

NOTES

1. McWhinney, Jim. "Getting Through the Medicare Part D Maze" January 9, 2006 (http://www.investopedia.com/articles/06/MedicarePartD.asp).

2. Pear, Robert and Edmund Andrews, "White House Says Congressional Estimate of New Medicare Costs Was Too Low," *New York Times*, national, February 2, 2004, A14.

3. Pear, Robert, "New White House Estimate Lifts Drug Benefit Cost $720 Billion," *New York Times*, late edition, February 9, 2005, A1.

4. Oliver, Thomas R., Philip R. Lee, and Helene L. Lipton, *Milbank Quarterly* 82, no. 2, 2004.

5. Budrys, Grace. 2005. *Our Unsystematic Health Care System*. 2nd edition. New York: Rowman and Littlefield Publishers, Inc., p. 168.

6. Ibid.

7. McWhinney, Jim. "Getting Through the Medicare Part D Maze" January 9, 2006 (http://www.investopedia.com/articles/06/MedicarePartD.asp).

8. Quadagno, Jill. 2005. One Nation Uninsured: Why the U.S. Has No National Health Insurance. Oxford University Press.

9. Weissert, William, "Medicare Rx: Just a Few of the Reasons Why It was So Difficult to Pass," *Public Policy and Aging Report* 13, 4 (2004): 3-4.

10. Budrys, Loc. cit.

11. Quadagno, Op. cit., p. 199.

Contributors

Berit Brogaard. Assistant Professor, Department of Philosophy, University of Missouri—St. Louis

Vern L. Bullough. Distinguished Professor Emeritus of History, Former Dean at SUNY College, Buffalo; Former Outstanding Professor, California University, Northridge; former professor, USC; fellow in the Medieval and Renaissance Center at UCLA; former Co-President, International Humanist & Ethical Union. *It is with deep sadness that we note the death, prior to this publication, of our friend and colleague Vern Bullough on June 21, 2006*

Carmela Epright. Associate Professor of Philosophy, Furman University; Clinical bioethicist and ethics consultant for several medical and psychiatric institutions in Greenville, S.C.

Faith Lagay. Senior Research Associate in the Ethics Standards Group at the American Medical Association

Kristy Maher. Associate Professor and Chair, Sociology Department, Furman University

Mason Olds. Professor Emeritus of Religion and Philosophy at Springfield College, former editor of the journal Religious Humanism (1990-1995); author *American Religious Humanism*; Adjunct Professor at both The Citadel and The College of Charleston

235

Howard Radest. Former Director [Headmaster], Ethical Culture Fieldston School, New York; Leader, American Ethical Union, Adjunct Professor of Philosophy, University of South Carolina, Beaufort, Co-Chair, Biomedical Ethics Committee, Hilton Head Regional Medical Center

Philip Regal. Professor of Ecology, Evolution and Behavior, University of Minnesota, Minneapolis

Andreas S. Rosenberg. Professor (emeritus) of Laboratory Medicine, Pathology, Biochemistry, and Biophysics, University of Minnesota, Minneapolis

Harvey Sarles. Professor of Cultural Studies and Comparative Literature, University of Minnesota, Minneapolis

David Schafer. Retired Molecular Biologist, Veterans Administration Hospital, West Haven, Connecticut; Founder of Humanist Association of Central Connecticut; President (2003-), HUUmanists (UUA Humanists).

Robert B. Tapp. Professor (emeritus) of Humanities, Religious Studies, and South Asian Studies, University of Minnesota, Minneapolis; Dean (emeritus) and Faculty Chair, The Humanist Institute, New York City

Stephen P. Weldon. Assistant Professor, History of Science, University of Oklahoma and Bibliographer, History of Science Society

Michael Werner. Sales & marketing, hi-tech adhesives; former President, American Humanist Association